INVENTING THE UNIVERSE

SUNY Series in Ancient Greek Philosophy
Anthony Preus, Editor

INVENTING THE UNIVERSE

Plato's *Timaeus*,
The Big Bang,
And the Problem of Scientific Knowledge

Luc BRISSON and F. Walter MEYERSTEIN

State University of New York Press

Published by
State University of New York Press, Albany

Printed in the United States of America

For information, address State University of New York Press,
State University Plaza, Albany, N.Y., 12246

Production by Marilyn P. Semerad
Marketing by Bernadette LaManna

Library of Congress Cataloging-in-Publication Data

Brisson, Luc.
    [Inventer l'univers. English]
    Inventing the universe : Plato's Timaeus, the big bang, and the
problem of scientific knowledge / Luc Brisson & F. Walter
Meyerstein.
       p. cm. — (SUNY series in ancient Greek philosophy)
    Includes bibliographical references and indexes.
    ISBN 0-7914-2691-2. — ISBN 0-7914-2692-0 (pbk.)
    1. Plato. Timaeus. 2. Cosmology, Ancient. 3. Big bang theory.
4. Cosmology. 5. Science—Philosophy. 6. Knowledge, Theory of.
I. Meyerstein, F. Walter. II. Title. III. Series.
B387.B7313 1995
113—dc20
                                    95-18054
                                        CIP

10   9   8   7   6   5   4   3   2   1

CONTENTS

ACKNOWLEDGEMENTS

This book is the authors' English revised version of their *Inventer l'Univers*, first published in Paris in 1991.

Translations of quotations from Plato's *Timaeus* are Cornford's and from Aristotle's *Posterior Analytics* are Tredennick's, with minor modifications by the authors.

Thanks are due to the following people: Michael Chase and Jeffrey Reid who helped with the English translation, and Anne Brisson who drew the Figures.

Introduction:
The Problem of
Scientific Knowledge

Knowing the universe is *inventing the universe*. This proposition, daring in an age like ours that has witnessed such a prodigious accumulation of scientific knowledge, seems surprising; moreover, the remarkable mathematical theory put forward by modern science to explain the universe appears to refute it out of hand. And yet, a careful examination of two fundamental cosmological models, Plato's *Timaeus* and the contemporary standard Big Bang model, shows that the kind of knowledge called "scientific" ultimately rests on a set of irreducible and indemonstrable formulas, pure inventions of the human mind, retained solely through recourse to this simple operative argument: "it works."

Confronted by this fact, philosophical investigation has wavered between two equally unpalatable solutions: 1) posing the "axiom that justifies all axioms," thus becoming meta-physical, non-rational or even mythical, or 2), after having lucidly established its inherently impassable limits, indefatigably driven by a desire to understand, by a nostalgia for the absolute, transgressing these same limits while unceasingly attempting to reinvent a simple and ordered universe.

The analysis of this perpetual renewing of the question, this unceasing quest, a veritable task of Sisyphus, constitutes the object of this book.

Situating the Problem: Our Essential Presuppositions

Previous to any analysis, a certain number of premises must be established,[1] the rules of the game must be agreed upon. The essential premises and definitions we require in order to formulate the problem of knowledge are the following:

1) The present analysis focuses on what in this book is called "scientific" knowledge. Knowledge acquired by other means or methods, such as knowledge acquired by way of divine revelation or mystical illumination, or even knowledge resulting from "common sense" or from habit is excluded from the analysis. We equally leave out the sheer accumulation of precise experimental data, important as it is for the scientific endeavor. We focus on science as knowing, not on science as doing. Our concern is directed exclusively at knowledge deemed to put forward the *explanation* of a fact or of a thing, the particular knowledge garnered by the hypothesis-deduction-testing procedure. Following in this Plato and Aristotle, we make a clear distinction between scientific knowledge (*episteme*) and mere opinion (*doxa*).[2]

This work thus deals exclusively with the type of knowledge which claims to propose, in a succinct and clear manner, a *causal explanation* to a determined set of phenomena that can be apprehended in the physical world, in nature. This scientific knowledge will be proved to be essentially *the list of axioms and rules of inference* supporting the formalized system in which consists the scientific explanatory theory put forward in each case. On the other hand, we do not inquire into what "is" the physical world where the explained phenomena are apprehended; we leave aside its ontological status. It suffices that such scientific explanations "work," that the scientific knowledge afforded by these axioms and by these rules of inference confers on mankind a considerable power over nature, the power to carry out, in certain cases, predictions with appropriate exactness. In short, our analysis deals with this question: what can be *said* about scientific knowledge?

2) Science explains *phenomena*, i.e., things and facts, *pragmata* according to Aristotle. These phenomena can be apprehended only by the senses and through a *measur*ing operation.[3] Of the five human senses, two serve especially this function: sight and touch. Consequently, when in the *Timaeus* he wishes to evoke the material world, Plato speaks of a "visible and tangible" (*horatos kai haptos*) world. But it turns out that "there is no science through sensation" (Aristotle, *Posterior Analytics* I 31, 87 b 28). Why? Because all that is material is found to be in time, nothing material can escape its never ceasing flow. Consequently, we are compelled to admit the crucial distinction between what "is" and what "becomes," a distinction which will prove fundamental when it comes to the elucidation of what can be said about this "scientific knowledge."

3) From the fact that every event is always sense-perceived at a definite place in space and at a definite instant in time, at some particular *hic et nunc*, follows the main characteristic of the material world: the material world never "is," it is always "becoming." In other words, the fundamental characteristic of the material world is *kinesis*, change, in the widest acceptation of the word. Philosophers and scientists however search for *necessary* explanations, for (universal) unchanging *laws* of nature, of which Newtonian gravitation is an example. It nevertheless turns out that between these two worlds, the material world of change and becoming and the world of being, of immutable Platonic forms or of mathematical theories, *no logical relationship can be established*; here lies a *hiatus irrationalis*!

4) This paradoxical situation is located at the center of our investigation. Thinkers from Plato to Einstein have been faced by this paradox: no logical relationship can ever be established between the formal axiomatic system, the idealized construction that we call a "scientific explanation," and the experimental

information supplied by our senses, by means of a measuring operation.[4] Einstein, in a letter to M. Solovine dated May 7, 1952, points out this absence of logical relationship between what he calls the set "S" of theoretically deduced propositions (*die gefolgerten Sätzte*) and the level "E" of sensible perception (*die unmittelbaren Sinnes-Erlebnisse*).[5]

It is precisely because the scientific method has shown itself to be so fruitful and powerful, especially in the twentieth century, that we propose to clarify the terms of the discussion that makes apparent a *hiatus irrationalis* between the intelligible and the sensible. And we intend to conduct our investigation by placing on an equal footing the results of the most recent scientific theories and those, still valid and fundamental, put forward by the Greek philosophers.

The Procedure Followed in this Analysis

This book is divided into three parts, each of which proposes a different approach to the problem of the *hiatus irrationalis*.

Part I: The Model of the Universe in Plato's Timaeus

Our interest in Plato's *Timaeus* is explained by the unique situation of this work in the history of Western philosophy.

1) The authenticity of the *Timaeus*, a work that has reached us assuredly complete after more than twenty-four centuries, is unquestionable. Through its Latin translations, it had an influence on Mediaeval philosophy and, after the rediscovery of its original text at the time of the Renaissance, it played a crucial role in the thought of scientists such as Kepler and Galileo.

2) For the first time in the history, Plato, in the *Timaeus*, submits the *problem of scientific knowledge* to a complete analysis, clearly stating that the character of necessity and ideality attributed by him

to a valid scientific explanation cannot follow in a non-mediated way from sense-perceived experiences.

3) To solve this problem, and again for the first time in history, Plato applies what will become "the" method of the scientific inquiry. This procedure requires as its first step the establishing of a list of premises or axioms. Once such a list has been drawn up, the scientist tries to verify if the propositions, deduced from these axioms according to certain precise rules of inference supposedly known and admitted, present a reasonable and adequate correspondence with sense-perceived experience.

4) Finally, and even more astonishingly, Plato, for the first time in history, uses mathematics as a tool to express and to deduce the consequences which derive from the postulated axioms. Indeed, starting with Aristotle, philosophers have, for their cosmological speculations, clearly recognized the scope and the power of the scientific method, that is to say, of the logical deduction of propositions starting from certain previously admitted hypotheses; but their tool remained common language. Although one can adduce several reasons to explain this situation, it is nevertheless surprising to note that, between Plato and Galileo, that is during twenty centuries, no one seems to have really appreciated the enormous power of mathematics for manipulating abstract concepts, particularly when establishing cosmological models.

Having put forward these points, we now explain the methodological procedure we will follow in the first part of this book. To begin with, we consider the *Timaeus*, and other works by Plato, not only as historical documents, but as texts whose complexity and consistency support an interpretation meeting up to date requirements; however, in so doing, we take all the necessary precautions to avoid introducing into these ancient texts manifestly foreign modern connotations. Consequently we consider that in the *Timaeus* Plato offers a particular "scientific" model to explain the universe, in accordance with the definition of science adopted in this book and with the obvious *caveat* that we are dealing with a cosmological model established in the fourth century BC.

In our analysis, we will endeavor to display the axiomatic structure which underlies the *Timaeus*. In fact, we will exhibit, as completely as possible, the list of axioms, or primordial assumptions, upon which Plato's cosmological system is ultimately founded. This procedure is interesting on two accounts: first, a relatively short list of axioms is all that Plato needs in order to propose a necessary and demonstrative scientific explanation of the ever changing phenomena observed in the material world. Second, in many respects this list of axioms is similar to the list of premises that contemporary science has also to admit; the common character shared by these assumptions will be brought to light by our analysis of contemporary Big Bang cosmology.

Finally, we will focus on the question raised by the experimental verification (better: falsification) of a scientific explanation, a question we will address on several occasions throughout this book. The *hiatus irrationalis* between scientific explanation (or scientific theory) and sense-perceived experience is so definitive that no non-mediated occurrence from the sense-perceptible world can modify the ontological and epistemological status of an idealized construction such as a formal axiomatic system. Experimentation is but a *method* allowing one to make a *choice* between different sets of axioms. Certainly, this method is of undeniable importance, considering that it allows the establishment of an empirical operative connection between an explanation (theory) and reality, the connection being expressed by "it works;" however, experimentation can never succeed in ascending from the particular to the universal. When Ptolemy added epicycles to his system in order "to save the phenomena" in the sky, the geocentric axiom was not questioned. But when Copernicus proposed a different set of axioms, the comparison and the *choice* between different astronomical *models* became possible. Let us remark furthermore that any question *about* the axioms is, in general, equivalent to the addition of a supplementary axiom, or, what amounts to the same, represents the transition to a metasystem. We will come back to this matter in the final part of this book.

On the other hand, the fact that Plato probably never thought of anything corresponding to an experiment, that he never conceived the controlled and repeatable observation of some particular physical arrangement depending only on a drastically reduced number of parameters, compared with the hypercomplexity found everywhere in the material world, is of no consequence, because the entire information content, all the "scientific knowledge" of the scientific explanation (theory), is enclosed in its axiom system. As a matter of fact, from this axiom system all the propositions susceptible of experimental verification can be deduced, and only the deduced theorems of the system permit the calculation of the predicted value of some physical variable that an experiment can measure and thereby either verify or falsify. In a word, one must carefully keep separate *scientific knowledge* from the scientific method of *choice*: the *scientific method of choice* is directly based on the technology of experimentation, whereas scientific knowledge is logically independent from this technology. To sum up, the scientific method allows a choice between competing axiom systems and, in general, the axiom system leading to more exact predictions is preferred.

Between this scientific method and scientific knowledge, which ultimately reduces to the list of axioms, a positive feedback relationship is established. The power of prediction which derives from this method is considerable, explaining why, even if this predictive power remains incomprehensible ("extra-logical" in Einstein's words), its success has established it as the method *par excellence*. Nevertheless, only the list of axioms and the rules of inference count for our analysis, since they contain all the relevant information a theory can convey.

However, Plato in the *Timaeus* does not admit, as scientific knowledge, a knowledge characterized by a purely functional definition of the sort: If a formal axiomatic system happens to be capable of producing theorems leading to sufficiently precise predictions, this system is accepted as "scientific," and no further inquiry shall be considered necessary. For reasons we will make clear, Plato refuses such a type of knowledge; or, more precisely, he would never qualify such an endeavor as *episteme*.[6] A purely operative definition of science, a definition of science which renounces the critical examination of its premises, is not only intellectually unappealing but at the same time abandons all quest for certitude, even when a remarkable correspondence with the empirical reality may eventually be uncovered.

Yet neither Plato nor the great majority of the philosophers who came after him succeeded in avoiding this impasse without appending to their theories a set of peculiar metaphysical entities, the necessity of which is hard, if not impossible, to prove from observation and experience. This will be documented by two precise examples: in the first part of the book, Plato's *Timaeus*; in the second part, the contemporary Big Bang model of cosmology.

Part II: The Big Bang Model of Cosmology

The detailed analysis of the *Timaeus* offered in the first part of this book will give an indication of what constitutes scientific knowledge for Plato. Plato supposes that a benevolent demiurge has framed, from a few primordial and simple mathematical building blocks, a universe in which all phenomena find their scientific explanation if and only if the scientist succeeds in uncovering the mathematical formula utilized by the demiurge for the assembly of the *kosmos*. In so doing, Plato hopes that the *hiatus irrationalis* could be bridged by mathematics, considered to be a purely rational construction. Furthermore, and since the demiurge is a benevolent one, he will guarantee, so far as possible, that the mathematics used to order the universe will be easy enough to become accessible to mankind. This metaphysics notwithstanding, Plato in the *Timaeus* asserts that all a human being can ever hope to know about a phenomenon is the mathematical formula expressing its change (*kinesis*). And, in spite of the twenty-four centuries separating them, contemporary cosmology seems to follow a course in more than one way parallel to Plato's endeavor.

But first this question: Why consider the Big Bang model as the scientific paradigm? Several reasons can be given: i) the Big Bang model is a cosmology and, in this sense, it is directly related to the *Timaeus*; ii) very few physical theories can claim such a wide scope or have attracted such a general interest; and, iii) from the standpoint of a philosophy of knowledge, science has nothing that can rival with a model purporting to describe the *entire* universe.

To remain methodologically consistent, we enumerate the axioms and assumptions indispensable for the construction of a coherent mathematical model of the universe, i.e., for the standard Big Bang model. In so doing, we will endeavor to show that it is possible to analyze this model as an ordered succession of premises reflecting the scientist's prejudices about space, time, causality, etc. And the set of these axioms will allow the construction of a universe where the mathematics we are familiar with can perform the duty of structuring an explanatory system. Once the fundamental characteristics of this model become clear, we will follow with a description of those observational facts deemed to verify the model, as well as the most important of those which seem to falsify it.

The main intention of this program is to highlight the purely intelligible character of the construction of the model. In fact, no *logical* connection can be established between a *N*-dimensional topological metric space and the sense-perceived "world". Furthermore, if we provide a rather technical description of the Big Bang model, it is to exhibit clearly the following aspects, which seem essential to our analysis.

1) A certain number of opposites, of which the pair being/becoming represents the fundamental antonymous couple of the *Timaeus*, can be found as well in the Big Bang model; this is shown in this Table:

Timaeus		Big Bang	
Being	Becoming	Model	Reality
Original	Copy	Theory	Observation
Eternity	Change	Pattern	Complexity
Order	Disorder	Prediction	Complexity
Symmetry	Chaos	Symmetry	Complexity

2) These opposites show once again the *hiatus irrationalis* to which we have repeatedly alluded. In the *Timaeus* as well as in contemporary cosmology, a central status is accorded to the concept of symmetry. In the *Timaeus*, the contrary of *symmetros* is *ametros*, an epithet qualifying the chaos which is devoid of measure, which is not calculable as we would put it today. In fact, this is the central idea developed in the *Timaeus*: the order established by the demiurge in the universe becomes manifest as the symmetry found at its most fundamental level, a symmetry which makes possible a mathematical description of such an universe. The homology between the Platonic and the modern approach deserves to be highlighted. Symmetry, a concept derived from geometry, is also the pivotal notion around which the Big Bang model will be constructed. Furthermore, the opposition symmetry/complexity will be used by Plato, as well as by modern cosmology, to distinguish in their modeling of the universe three distinct domains:

• The very large, the astronomical, the realm of the celestial bodies, of galaxies and of clusters of galaxies. In this domain, according to the description provided by both models, everything is symmetric, homogeneous, isotropic, simple. This is the realm of eternal circular motions, of mathematical physics.

• The very small, the microscopic, the realm of the elementary building blocks. Here also, the models postulate symmetry and simplicity. This is the domain of the perfect Platonic polyhedra and of their mathematical interactions, the realm of the physics of elementary mathematical particles.

• The sublunary world, which includes everything mankind is able to resolve, or analyze, into its elementary constituents. This is the universe of distinct galaxies, of people, of viruses. It is the world of the particular, of individuals. It is the realm of the complex.

3) For both models the following crucial axiom must be admitted. The entire universe is reducible to some simple mathematical primordial elements, and everything happening in the universe, every time-dependent change (*kinesis*), is also reducible to simple[7] mathematical interactions between simple mathematical elements. The result will be an ordered world, a *kosmos*. And this order is the consequence of the symmetries introduced axiomatically in the description of the universe.

4) Finally, contemporary cosmology adopts an attitude very similar to that of Plato's, in that the discovering of the mathematical formula "that works" is the only objective for scientific research. But once a theory has been found, its axioms and its inference rules having been selected as those that "work" better than other models, nothing further can be "scientifically" said about the theory itself.

This limited knowledge is certainly not what Plato considered to be *episteme*. In Plato's thought, *episteme* can only refer to the particular kind of knowledge that has as its sole object the intelligible Forms. This is *true knowledge* (*alethes logos*), radically superior in his eyes to a simple enunciation of a functional theory. But true knowledge is only attainable by the gods, and a few selected men: the philosophers. It follows that the mathematical model devised by common mortals can never be more than a verisimilar account of reality, an *eikos logos*, a plausible copy of true reality. Modern science shares this predicament: every theory is at best a provisional explanation of some phenomenon, which remains open-ended, since a theory can only resist falsification, but can never be verified.

Thus the *Timaeus* and the Big Bang model share a limited optimism about what can be attained as scientific knowledge. Based on the postulate proclaiming the universe to be ordered and reducible to a few simple elements interacting in a simple mathematical way, the scientist arrives only at a *verisimilar* image of *episteme*, he can offer but a copy of reality, hoping that the progress of his endeavor will lead his knowledge ever closer to actual reality. But even this unpretentious ideal seems to be severely constrained if the conclusions of the recently developed Algorithmic Information Theory are to be taken into account. In order to complete our analysis, we shall use this powerful tool developed by Gregory Chaitin.

Part III: Algorithmic Information Theory

Algorithmic Information Theory is so recent that some of the conclusions we derive from it may appear highly paradoxical. We nonetheless justify our special interest in this theory by the following considerations:

1) As the examples of the first two parts should prove, the acquisition of scientific knowledge requires a previous intelligible operation: the physical situation under consideration must be expressed or formulated in a symbolic language, a language based on pure symbols, unrelated to any physical substratum whatsoever (for instance: a N-dimensional topological space, a symmetry group, a quantum number, etc., or more modestly, but similarly, a

distance in meters, an angle in degrees). This is the *preliminary scientific description* of the phenomenon under consideration. As shown by N. Wiener and C. Shannon, such description can then always be coded in binary language, as a sequence of binary symbols such as 0 and 1. The same remark applies to any scientific theory supposed to "explain" the change (*kinesis*) observed in the particular physical situation selected.

Algorithmic Information Theory allows the mathematical demonstration of significant theorems pertaining to such abstract sequences of only 0s and 1s. Such sequences, as will be shown, can represent what previously has been called "scientific knowledge."[8] But more fundamental is the fact that the theory allows us to measure the complexity of a binary sequence. In so far as a scientific theory is a representation of reality, and in so far as we can speak about such a theory, it becomes feasible to envision this representation as a binary sequence of a certain length and complexity. At this point, questions about the complexity of a scientific theory, or its information content, become pertinent, and a much more discerning and perceptive approach to the problem of a scientific knowledge, of its *hiatus irrationalis*, becomes feasible.

2) To reformulate the problem of scientific knowledge in the terms of theorems demonstrated by the Algorithmic Information Theory may seem odd, but the conclusions reached are even more surprising. We will *prove* that the probability for our universe being ordered according to some simple mathematics, and reducible to some simple elementary components, is infinitely small, except if a benevolent god has decided otherwise. We will also *prove* that all knowledge a scientific theory, such as the one developed by the *Timaeus* or by contemporary cosmology, can convey, reduces to the bare enunciation of such a theory, that is, to the listing of the basic set of its axiomatic propositions and of its admitted inference rules. Therefore we will *prove* that this constitutes a definite *limit* in two senses: anything surpassing the complexity of the theory is undecidable *in* the theory; any *question* requires additional axioms.

But should the theory be the "final" theory about which some physicists dream, then it is itself totally ineffable, no *rational* discourse about such a final theory is possible.

3) Not unexpectedly, this new approach is capable of expressing, but this time in a rigorous mathematical manner, ideas Aristotle had already developed mainly in the *Posterior Analytics*. In fact, our application of the Algorithmic Information Theory will show that Aristotle's ideas, now expressed in the sophisticated language of contemporary mathematics, have lost nothing of their interest.

NOTES

1. As Aristotle (*Posterior Analytics* I 1, 71 a 1-2) has expressed it, all knowledge acquired "by way of argument *(dianoetike)* proceeds from pre-existent knowledge."

2. "We suppose ourselves to possess unqualified scientific knowledge of a thing, or of a fact *(pragma)* as opposed to knowing it in the accidental way in which the sophist knows, when we think that we know the cause on which the fact depends, as the cause of that fact and of no other, and, further, that the fact could not be other than it is." Aristotle, *Posterior Analytics* I 2, 71 b 9-12.

3. Actually, no unmediated sense-perception can become a measure. A "measure" already presupposes a theory. We here skip this complex problem in order to concentrate our investigation exclusively on what we have defined as "scientific knowledge."

4. This lack of relationship caused E. Wigner to write his famous article "The Unreasonable Effectiveness of Mathematics in the Natural Sciences," *Communications on Pure and Applied Mathematics* 13, 1960, p. 2, where one can read: "The enormous usefulness of mathematics in the natural sciences is something bordering on the mysterious and there is no rational explanation for it."

5. Concerning how science establishes a relation between the two spheres, he writes: "Diese Procedur gehört genau betrachtet ebenfalls der extra-logischen

Sphäre an, weil die Beziehung der in den *S* auftretenden Begriffe zu den Erlebnissen *E* nicht logischer Natur sind." And he concludes: "Die Quintessenz ist der ewig problematische Zusammenhang alles Gedanklichen mit dem Erlebbaren (Sinnes-Erlebnisse)" (Facsimile of this letter in Einstein, *A Centenary Volume*, A.P. French ed., London 1979, p. 271): "This procedure, looking more closely, belongs as well to the extra-logical (intuitive) sphere because the relation between the concepts intervening in *S* and the experiences *E* are not of a logical nature. The quintessence is the relationship of the eternally problematical connection between what is thought and what is experienced (through the senses)."

6. As Wigner says in the article already quoted: "... we do not know why our theories work so well. Hence their accuracy may not prove their truth and consistency." (*op. cit.*, p. 14)

7. "Simple" here means: accessible in principle to human investigators. Simplicity, a rather subjective, psychological concept, plays nevertheless a fundamental role in cosmology; complexity, its anathema, can, on the contrary, be precisely defined.

8. Every scientific theory must be communicable, must be formulated in some inter-subjective language. Solipsistic "knowledge" is certainly not "scientific" knowledge.

Part I:

The Model of the Universe in the Timaeus

The aim of the *Timaeus* is multiple, comprising a cosmological model and simultaneously describing the origin of mankind and the constitution of an ideal city which is both the critical reverse of the Athens in which Plato lived and the model it should follow. It is therefore impossible to separate neatly in the *Timaeus* that which pertains to cosmology and that which is dependent on other areas of knowledge: mathematics, physics, chemistry, biology, medicine, psychology, sociology, politics and even religion. All this is tied together in one dense web. None of these areas of learning displays the real autonomy that they have been able to gain two millennia later. A reading of the *Timaeus* can therefore only avoid anachronism if it recognizes from the outset this lack of autonomy. Nonetheless, the *Timaeus* is a work on cosmology, since it advances a model of the physical universe; it is even the first such work to have reached us in its entirety. What is more, for the first time, a model of the universe is proposed that professes to be totally mathematical.

A further difficulty in the *Timaeus* ensues from the intertwining of mythical narrative and scientific approach, a problem indissociable from the status of a discourse on the origin of the sensible world. To recall the origin of the sensible world is to describe *the coming into being of sensible reality*, which, by definition so to speak, no human being can ever have experienced. The philosopher who commits himself to this undertaking is as unprepared as the poet, Hesiod for example, who, in the *Theogony*, must have recourse to the Muses. And Plato must first set forth his own fundamental epistemology.

The Date of Composition of the Timaeus *and Its Dramatic Situation*

The *Timaeus* appears to follow the *Republic,* or a dialogue resembling this discussion of Justice, and it is followed by the *Critias,* an unfinished dialogue meant to expand upon a *Hermocrates* that was never written. It is essentially a discussion between four characters: Socrates, Hermocrates, Critias, and Timaeus.

The project of this discussion as a whole, within which the *Timaeus* is embedded, is above all political and tackles this question: how can the Athenians be reformed? The answer: by reminding them of their history (which Hermocrates would have done), and by evoking a distant past (which Critias does) where the organization of their City conformed to the ideal described by Socrates in the *Republic*. To show that this ideal is realizable on earth, to found this political project in nature, it is necessary to go back to the origin of man and to the origin of the world, and to explain how man, this microcosm, finds his place within the macrocosm, the universe which is but the sensible image of an intelligible model. Plato puts into the mouth of Timaeus, a citizen of Locri (in southern Italy) the long monologue wherein this ambitious program is set forth in detail.

On what date is the discussion between Socrates, Hermocrates, Critias and Timaeus supposed to have taken place? If we leave aside the problem presented by the relationship between this discussion and the one recounted in the *Republic*, the action must be situated between 430 and 425 BC. Socrates would then have been forty to forty-five years old.

It seems that the *Timaeus* and the *Critias* were written by Plato ten or twelve years before his death, between 358 and 356 BC. Translated into Latin, at least in part, by Cicero (106-43 BC) and by C(h)alcidius (fourth cent. AD), commented by many Platonists, including Proclus (fifth cent. AD), the *Timaeus* has reached us through papyri and manuscripts, the oldest of which was produced in Constantinople and goes back to the end of the ninth century of our era (more than a millennium after Plato's death).

The First Twelve Axioms of the Cosmological Model Advanced in the *Timaeus*

Fundamental to our analysis in this book is the following assumption: the cosmological model advanced by Plato is a "scientific model," in the strong contemporary sense of the term. It thus ensues that this construction, this theoretical model of the universe, must be assembled as a formal axiomatic system: a set of primordial propositions – the axioms – must first be established, and all observable manifestations of the universe must then in principle be deducible as theorems from these axioms.

To be sure, Plato may not have been perfectly conscious of the fact that he was constructing a model based on a list of axioms and that these axioms have no other justification than their epistemological consistency. But he accepted this idea: whoever finds a different set of axioms, whoever could advance another model or attempt a different explanation, is equally entitled to be listened to; and only the model best fitting the data, the model that "works best," will be chosen.

> ... but if anyone should put the matter to the test and discover that it is not so, the prize is his with all good will. (*Timaeus* 54a-b)

> ... but another, looking to other considerations, will judge differently. (*Timaeus* 55d)

This truly modern feature of Plato's cosmology deserves to be emphasized.

Axiom T1

Reality is separated into two domains: the intelligible Forms (*eidos, idea*), pure, eternal, immutable and simple; and the complex sensible particulars, ever-changing (*kinetos*) in time.

This separation of reality into two domains corresponds to the distinction between being and becoming, between that which remains forever identical and that which never ceases becoming different, between that to which the predicate "true" can be attributed and that to which this predicate is refused.

In the Platonic system, the intelligible Forms, generally called "ideas,"[1] are postulated as metaphysical entities, indispensable for explaining the world perceived by the senses. There is no veritable reality in the permanent change of all that is in time, of all that is always becoming, of all that we moderns call "material." Time is tied to change, but does not affect the eternal realities, which are timeless and changeless. As we can read in the *Timaeus*:

We must, then, in my judgment, first make this distinction: what is that which is always real and has no becoming, and what is that which is always becoming and is never real? That which is apprehensible by thought with a rational account is the thing that is always unchangeably real; whereas that which is the object of belief together with unreasoning sensation is the thing that becomes and passes away, but never has real being. (*Timaeus* 27d-28a)

Only the knowledge of intelligible Forms can really be said to be "true." But finite, ephemeral and limited human beings cannot in this world attain such a knowledge. The knowledge of these divine entities is the exclusive privilege of the gods and of a small number of their friends (*Phaedrus* 278d). Knowledge whose object is the world of sensible particulars, which Plato qualifies as "opinion" (*doxa*), is an inferior type of knowledge, since it can at best attain only verisimilitude. Here lies the root of the epistemological problem Plato tries to solve in the *Timaeus: how to know truly the sensible world which is ever-changing, whereas true knowledge (*episteme) can have as its objects only the intelligible Forms which moreover remain inaccessible to human beings?*

Axiom T2

The Good occupies a singular situation among the Forms.

The Good is one intelligible Form among others: Justice, Unity, Man, Animal, etc., but it plays a crucial role within the Platonic system, particularly in the *Timaeus*.[2] This Form confers upon the other intelligible Forms these distinctive features: beauty, harmony, order, simplicity. According to the cosmological model advanced by Plato, the other intelligible Forms then communicate these features to the sensible particulars.

Axiom T3

In the sensible world, all that becomes becomes as the result of a cause.

Causality only finds an application in the sensible world, since in the realm of intelligible Forms, which, following axiom T1, are eternal and immutable, there is no change, and, consequently, there is no cause-effect connection in the intelligible world. And in the sensible world, *changes* amount to the *relationships* between elementary components (cf. axiom T18 *infra*). These elementary components are eternal and immutable (cf. axiom T13 to T18 *infra*). But all change obeys causality, according to axiom T3. Any change of relationship will therefore always be the effect of another change of relationship, antecedent in rank or in time. By further postulating that inter-connections between changes, between relationships, can as far as possible be expressed mathematically (cf. axiom T12 *infra*), Plato constructs (invents, according to our terminology) a causally ordered mathematical universe.

The word "cause" (*aitia*), a term borrowed from the judicial vocabulary where it designated responsibility,[3] is used to designate this chain of relationships.

Axiom T4

The sensible world is the result of the ordering effort of a god.

One of the causes of the sensible world is a god, also called "father," "maker" and "demiurge," this last term being the most frequently used. This god does not *create* the world; his action is limited to the partial *ordering* of a primordial chaotic substrate (cf. axiom T7).

Axiom T5

The demiurge is good (*agathos*).

The goodness of the demiurge imposes upon him a certain way of acting (cf. axiom T10).

Axiom T6

The demiurge is not omnipotent.

This god is not omnipotent, for two reasons posited as axioms. The intelligible Forms and the *khora*, the primordial stuff or the "spatial medium," exist independently of him (axioms T1 & T7). 2) And the demiurge must face *anagke* (axiom T8) which always resists his ordering effort.[4]

The Platonic demiurge is a peculiar divinity; after his ordering effort, he retires from the universe (*Timaeus* 42e). This cosmological approach presents a radically "materialis" character. Mankind is left alone in a material, ever-changing world where divine intervention subsists only in the form of an imperfect, partial, mathematical order.

Axiom T7

The demiurge orders a primordial stuff, the *khora*.

Khora is at the same time that in which sensible particulars are found, i. e., space or place, and that of what they are made, i. e., something approximating matter. We translate *khora* as "spatial medium."

Khora is a hybrid entity. It is eternal, it exists even before the demiurge introduces, insofar as this is possible, order into it. But all that is found "in" the *khora*, and all that is produced "from" it, the sensible world, is ever-changing.

Plato acknowledged the difficulty in conceiving this "spatial medium;" and he was probably aware of the fact that he could not find a solution to the space/matter relationship; a problem that had to wait until the twentieth century to be solved. In fact, Plato asserts that we can only manage to conceive the *khora* by a sort of "bastard reasoning."

> ... *khora* is everlasting, not admitting destruction; providing a situation for all things that come into being, but itself apprehended without the senses by a sort of "bastard reasoning," and hardly an object of belief.

This, indeed, is that which we look upon as in a dream and say that anything that is must needs be in some place and occupy some room, and that what is not somewhere in earth or heaven is nothing. (*Timaeus* 52b)

... *khora* is the receptacle and the nurse of all becoming. (*Timaeus* 49a)

These quotations show how the spatial medium is at the same time "that in which" and "that from which" the sensible world is made.

Axiom T8

A cause, called *anagke*, perpetually resists the order which the demiurge attempts to introduce in the world.

The term *anagke* in ancient Greek is generally translated as "necessity." But the way Plato uses the term *anagke* in the *Timaeus* refers to a very different meaning from the one intuitively given to "necessity": constraint regarded as a law prevailing through the material universe. Plato holds *anagke* to be a "cause," but a negative one, qualified as an "errant cause *(planomene aitia)*" (*Timaeus* 48 a), since it represents a non-rational element permanently resisting the ordering effort of the demiurge (cf. axioms T9, T10, T11, T12).

Anagke is indeed an inherent property of the *khora* postulated in axiom T7. The effect of *anagke* is that, in the *khora*, before the demiurge's interventions, the four elements that are supposed to make up all of the sensible world (axiom T9) "behave without reason or measure *(alogos kai ametros)"* (*Timaeus* 53a). Nothing in the *Timaeus* allows us to know to what extent the demiurge, who is not omnipotent (axiom T6) has succeeded in imposing order on the universe. *Anagke* continues to manifest itself in the sensible world as an "errant cause," after the demiurge retires from the world. As a result, a factor of complexity and disorder always subsists in the universe.

It is easy to shrug off *anagke* as a myth. However, to claim that the *entire* universe must submit to simple mathematical rules accessible to mankind is at least as mythical.

Axiom T9

> **Sensible particulars, including heavenly bodies, are made out of four elements only: fire, air, water and earth.**

Here Plato follows the tradition inaugurated probably by Empedocles, and which was universally accepted until the birth of modern chemistry in the eighteenth century.

Axiom T10

> **All that a benevolent demiurge (axiom T5) endeavoring to introduce some order into the *khora* (axiom T4) can do is to use as his model a "perfect paradigm" and to attempt to bring it about that the result of his efforts be the best possible *copy* (*eikon*) of that model (axiom T6).**

In the *Timaeus*, one can read:

> Let us, then, state for what reason becoming and this universe were ordered by him who ordered them. He was good ... Desiring, then, that all things should be good and, so far as might be, nothing imperfect, the god took over all that is visible – not at rest, but in discordant and unordered motion – and brought it from disorder into order, since he judged that order was in every way the better. Now it was not, nor can it ever be, permitted that the work of the supremely good should be anything but that which is best. (*Timaeus* 29e-30b)

> Now whenever the maker of anything looks to that which is always unchanging and uses a model of that description in fashioning the form and quality of his work, all that he thus accomplishes must be good.[5] (*Timaeus* 28a-b)

Plato's epistemology postulates that the knowledge leading to truth *(episteme)* can only be that knowledge whose objects are the intelligible

Forms. This knowledge remains inaccessible to human beings living in the sensible world. They can only know the partial order, that imprint of the divine, which the demiurge attempts, insofar as possible, to introduce into the world. By reason of axiom T1, reality has been separated into two unbridgeable domains; now the mathematical order imposed by the demiurge provides a means to fill this gap; it represents the epistemologically necessary participation of sensible particulars in the Forms.

> ... [in the *khora* before the demiurge's intervention], these things were in disorder and the demiurge introduced into them all every kind of measure in every respect in which it was possible for each one to be in harmonious proportion (*analoga kai summetra*) both with itself and with all the rest. For at first they were without any such proportion save by mere chance, nor was there anything deserving to be called by the names we now use – fire, water, and the rest... (*Timaeus* 69b-c)

It follows from the axioms postulated thus far, that the effort of the demiurge consists in transforming, as far as possible, a chaotic, spatio-material substratum into a construction ordered according to symmetry, an operation that, by itself, will enable the *naming* of the things of the sensible world. In order to do this, the demiurge takes as a model what Plato calls the "perfect paradigm," the intelligible realm ruled by the Good. The universe, thus modeled, presents, as much as possible, beauty, symmetry, order, harmony, simplicity, etc.

Axiom T11

> **As a copy of a perfect paradigm, the sensible world made by the demiurge can be nothing other than a living thing whose body is made from the four elements (cf. axiom T9) and whose soul (*psukhe*) is endowed with reason (*nous*), (cf. axiom T12).**

According to axiom T1, the sensible world undergoes perpetual change. And according to axiom T10, the demiurge introduces partial order into this change. Now the major cosmological problem for ancient Greeks was to account for what is partially ordered in the sensible world: the engendering of man by man, the ordered succession of the seasons, and above all, the most regular movement that can be observed, that of the heavenly bodies. Plato explains these changes by assimilating the sensible world to a living being. The distinctive feature of a living being is the *autonomous* principle of *ordered* change and movement (*kinesis*). This autonomous source of order was called "soul (= *psukhe*)" by Plato, who here again gave a new meaning to an ancient word. Consequently, *if* any kind of scientific knowledge of the sensible world is to be obtained, that world must be presupposed to be equally endowed by an autonomous principle of ordered change and movement; it must be an ensouled (living) entity. Since the demiurge is good (axiom T5), he tries to endow the sensible world with the best possible soul, a soul blessed with reason (*nous*). And it is this soul endowed with reason that *directly* explains the regular, ordered and permanent, that is to say "rational," movement of the heavenly bodies.

> Taking thought, therefore, he [= the demiurge] found that, among things that are by nature visible, no work that is without reason will ever be better than one that has reason, when each is taken as a whole, and moreover that reason cannot be present in anything apart from soul. In virtue of this reasoning, when he ordered the universe, he fashioned reason within soul and soul within body, to the end that the work he accomplished might by nature be as excellent and perfect as possible. This, then, is how we must say ... that this world came to be, by the god's providence in very truth a living creature with soul and reason. (*Timaeus* 30b-c)

Since the world possesses a rational soul, and since order, which is an aspect of the Good, is infinitely superior to disorder, the change that affects sensible particulars will be ordered there where the world soul imposes itself like a "mistress and governor." (*Timaeus* 34c)

Plato postulates that perfect knowledge must have as its object perfect being, and since the sensible world is just a copy inferior to its model, because the demiurge is not omnipotent (axiom T6), he produces only a copy of a perfect paradigm (axiom T10). Therefore, the knowledge of the universe framed by such a demiurge must remain imperfect, amounting at best to a verisimilar account, to an *eikos logos*.

Axiom T12

As the vehicle of reason, the world soul is mathematically structured.

In the cosmology of the *Timaeus*, this axiom holds the place of the key epistemological axiom, since it secures the *a priori* possibility for any human knowledge of the universe. In Plato's epistemology, real knowledge (*episteme*) has no other objects than the intelligible Forms; it thus remains beyond the reach of human beings. The verisimilar account (*eikos logos*) humanity can attain is thus dependent on this astonishing hypothesis: the characteristics that the Good dispenses to the intelligible Forms, such as beauty, symmetry, order, harmony, proportion, etc. can be expressed in mathematical terms. Given the state of mathematics in Plato's time, this hypothesis is truly amazing.

The scope of axiom T12 can only be understood if we refer to the role *symmetry* plays in Plato's cosmology. In ancient Greek, *summetros*, a word made up of *metron* "measure" as second term, means properly "with a common measure." If things have a common measure, they are called "commensurable;" otherwise they are incommensurable. A common measure allows proportion to appear. Thus things may be said to be in due proportion (*analogoi*); and if the proportion remains always the same, these things are symmetrical as such. If such ideas are accepted, "symmetrical" can become synonymous with "harmonious" and even more importantly with "ordered" (= *kosmetos*).

In its more general acceptation, symmetry describes those aspects of a thing that remain unchanged, if that thing is considered from different points of view, for example the left side and the right side of a face, or if that thing undergoes certain transformations, for example a parallel translation or a rotation around an axis (the circle and the sphere playing here a crucial role).

The essential aspect of symmetry is the following: some properties of a thing remain invariable throughout change, something remains "analogous," because proportion (= *analogia* in Greek, *ratio* in Latin) is preserved. But only the intelligible Forms, sole objects of *episteme,* are, according to Plato, unchangeable, immutable and eternal. Plato's aim, in the *Timaeus,* is to propose a cosmology which secures a verisimilar account (*eikos logos*) of the universe, since a true account (*alethes logos*) of it is out of reach. In our ever-changing sensible world, a verisimilar account, a partial knowledge, is possible if and only if this world shares the immutability of the intelligible Forms, even if only partially. This Plato assumes as the key tenet of his cosmology. Consequently, the aim of cosmological inquiry is to uncover such invariable properties. Plato daringly extrapolates: these reflections of the divine are found in the world as symmetry, and symmetry can only be grasped mathematically. In this way, the entire program for science is outlined; it has not changed since.

Plato's extrapolation relied on an important discovery made in his time. The sound of a musical string instrument remains consonant when the lengths of the strings are changed, if the mathematical proportion (*analogia*) between the lengths of the strings remains unchanged. The Pythagoreans had discovered that what remains unchanged is a mathematical formula giving, in terms of integers, the proportion between the lengths of the strings.

One can thus predict that two different musical instruments, such as *A* and *C,* will identically produce a consonant harmonious music, if the lengths of their strings are in the correct mathematical proportion *B.* The participation of the sensible (material strings) in the intelligible (musical proportion) is thus mediated by mathematics. Plato will bestow this Janus-like characteristic of mathematics upon the world soul. Thereby mathematically expressible symmetry is promoted to the rank of the necessary *a priori* condition for any scientific knowledge of the sensible world. Plato puts forward one of the key tenets of contemporary physics, but the mythical language he employs has obscured this fact for centuries.

The World Soul in the *Timaeus*

The world soul is that entity which mathematically orders the universe, for it is destined by the demiurge to rule over the universe "as its mistress and governor." (*Timaeus* 34c) The mathematical order that governs the universe is determined by two characteristics of the world soul. On one hand, it participates both in being and becoming; by this means, it appears as the mediating agent between the eternal Forms and a sensible world that never ceases changing (axiom T1). Furthermore, the world soul exhibits a mathematical structure; and a mathematical order becomes manifest only there where the soul rules. In the Platonic system, the soul in general, and the world soul in particular, are the causes of ordered movement (axiom T11). In a universe governed by a mathematically structured soul, all change is necessarily governed, as far as possible, by mathematical precepts. Following this, that which at first seemed to exceed all possibility of rational analysis – reality exclusively perceived by the senses foreign to reason, and which never ceases changing – becomes amenable to a verisimilar account.

The demiurge achieves his objective as follows. In the first place, he introduces a soul into the world's body, which, being a copy of a perfect original, is endowed with the most perfect form, and thus presents the aspect of a gigantic sphere, the sphere being the most symmetrical figure. Moreover, to avoid infinite regress, the world soul must be autonomous, that is, it must be the cause of its own movements. As a result, the soul is a complex construction displaying two remarkable features: its movements are circular, circular movement being the most symmetrical movement, and they obey the laws of musical harmony, because musical harmony can be considered an aspect of the Good (cf. the "music of the heavenly spheres"). As well as being circular, the movements of the world soul maintain permanently a constant speed, a property which gives them the most perfect symmetry. Since immutability means perfection, this permanence and regularity, which are ultimately of a mathematical nature, allow the sensible world, where all movements are governed by the movement of the world soul, to partake, to a certain extent, in the eternity and the stability of the intelligible.

The Substance of the World Soul

The description of the world soul takes the form of a mythical narrative recording what the demiurge does. The intervention of the demiurge does not however violate the postulate formulated in the

Phaedrus (245c-246a) concerning the unengendered character of all principles, because this intervention does not imply an origin in time. It reveals two things: the soul is ontologically dependent upon the intelligible Forms, and furthermore is a reality intermediate between intelligible Forms and sensible particulars.

To compose the most fundamental entity of his cosmology, Plato makes use of the three most general notions in his metaphysical system: Being, Sameness and Difference. All reality comprises these constituent elements as described in the *Sophist* (254d-259b). All reality "is," the first requirement in metaphysics. Considered in its relationship with all that it "is not," this Being maintains its identity, which brings about the second fundamental concept, Sameness. But this Being only maintains its identity, because it is different from all that is not itself, because it is Different from all the rest (a horse *is*, it is a *horse* and it is *nothing else*, for example, a cat).

Furthermore, as the world soul must play the role of mediating agent between the sensible and the intelligible, its constituent elements are situated in it on an intermediary level between indivisibility, which characterizes the intelligible, and the divisibility, which characterizes the sensible. This is what the demiurge tries to bring about in performing these mixtures:

> Between the indivisible Being that is ever in the same state and the divisible Being that becomes in bodies (axiom T1), he compounded a third form of Being composed of both. Again, in the case of Sameness and in that of Difference, he also on the same principle made a compound intermediate between that kind of them which is indivisible and the kind that is divisible in bodies. Then, taking the three, he blended them all into a unity, forcing the nature of Difference, hard as it was to mingle, into union with Sameness, and mixing them together with Being. (*Timaeus* 35a-b)

In this difficult passage, illustrated in the following diagram, Plato expresses two ideas: i) the soul comprises the same constituent elements as any other reality: Being, Sameness, Difference; and ii) it is an intermediary reality between the intelligible and the sensible.

FIRST MIXTURE   SECOND MIXTURE

indivisible Being
                     } intermediate Being
divisible Being

indivisible Sameness
                     } intermediate Sameness   } World Soul
divisible Sameness

indivisible Difference
                     } intermediate Difference
divisible Difference

From these mixtures, a concoction results which Timaeus describes as a mass of metal. Solidified, this mass of metal serves the demiurge as material for constructing the armillary sphere which displays the motor function of the world soul, on which all the movements of the universe, including those of the planets and the fixed stars, depend.

The Mathematical Structure of the World Soul

The world soul is framed as an armillary sphere, since, although it is supposed to be the *principle* of movement of the heavenly bodies as well as the *principle* of changes within the whole universe, its primordial purpose is to account precisely for the observed movements of the heavenly bodies, and allow a mathematical description of astronomical phenomena.

The movements of the heavenly bodies seem to present two characteristics: *permanence* and *regularity*, characteristics known from the remotest antiquity and which led mankind to regard these bodies as divine beings, as opposed to the hypercomplex sublunary realities, subject to movements apparently devoid of all regularity. To account for these two characteristics, Plato formulates two postulates, flowing from axioms T10, T11 and T12. 1) Movements of the heavenly bodies are circular, thus they are *permanent*; 2) Movements of the heavenly bodies obey the laws of the various types of the mathematical proportions known at the time; thus, in spite of appearances, they are perfectly *regular,* i.e., mathematical.

The demiurge, whose actions are described in terms generally applied to a blacksmith, laminates the mass resulting from the mixture described above, and transforms it into a sheet. He begins by cutting this sheet lengthwise into two bands, which he somewhat paradoxically calls the band of the "Same" and the band of the "Different" even though these two bands are composed of the same mixture of Being, Same and Different. This operation accounts for the observed dissimilarity between fixed stars and planets. Next, the band of the Different is divided by the demiurge into seven sections to explain the movement of the "planets" known at that time. The apparently erratic (*planetes*) movement of the planets probably explains the name Different given to this band, to oppose it to the band of the Same which represents the apparently regular movement of the fixed stars.

Figure 1.1

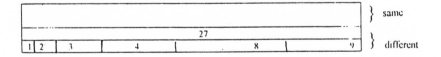

This first operation is not sufficient. It has allowed the formation of two bands. But these bands must then be bent to become those circles on which the heavenly bodies will move with the *permanence* provided by the perfect symmetry of the circle.

The *regularity* of the heavenly bodies' movement must still be accounted for: this is where proportion comes into play.[6]

The band of the "Different" is therefore divided into seven parts, according to the following series of integers: 1, 2, 3, 4, 9, 8, 27. It may be observed that this series corresponds to a double geometrical progression expressed in powers of 2 and 3 respectively:

$$2^0 \ 2^1 \ 2^2 \ 2^3$$
$$3^0 \ 3^1 \ 3^2 \ 3^3$$

But the mathematical explanation goes much further.

These seven numbers represent, as we will see further on, the orbital radius of each of the seven planets that gravitate around the earth. The earth remains immobile in the center, and the number 1 corresponds to the distance from Earth to Moon. Between these seven numbers, two series of proportional means are now inserted.

1) harmonic means
$$(x - a)/(b - x) = a/b; \text{ or } x = 2ab/(a + b)$$
2) arithmetical means
$$(x - a) = (b - x) \text{ or } x = (a + b)/2$$

Which produces:

1) resulting from the insertion of the harmonic and arithmetical proportional means in the first geometrical progression:

		harmonic means	arithmetical means
a=1	b=2	4/3	3/2
a=2	b=4	8/3	3
a=4	b=8	16/3	6

that is to say: 1, 4/3, 3/2, 2, 8/3, 3, 4, 16/3, 6, 8

2) and as results of the insertion of the harmonic and arithmetical proportional means in the second geometrical progression:

	harmonic means		arithmetical means
a=1	b=3	3/2	2
a=3	b=9	9/2	6
a=9	b=27	27/2	18

that is to say: 1, 3/2, 2, 3, 9/2, 6, 9, 27/2, 18, 27

Furthermore, if we consider this double series of results, we notice that, between the harmonic and arithmetical proportional means in each of the two geometrical progressions given at the outset, there exist only three types of interval: 4/3, 9/8, 3/2.

$$1 \quad 4/3 \quad 3/2 \quad 2 \quad 8/3 \quad 3 \quad 4 \quad 16/3 \quad 6 \quad 8$$
$$\underbrace{}_{4/3} \underbrace{}_{9/8} \underbrace{}_{4/3} \underbrace{}_{4/3} \underbrace{}_{9/8} \underbrace{}_{4/3} \underbrace{}_{4/3} \underbrace{}_{9/8} \underbrace{}_{4/3}$$

$$1 \quad 3/2 \quad 2 \quad 3 \quad 9/2 \quad 6 \quad 9 \quad 27/2 \quad 18 \quad 27$$
$$\underbrace{}_{3/2} \underbrace{}_{4/3} \underbrace{}_{3/2} \underbrace{}_{3/2} \underbrace{}_{4/3} \underbrace{}_{3/2} \underbrace{}_{3/2} \underbrace{}_{4/3} \underbrace{}_{3/2}$$

These three types of interval correspond to the musical relationships already known in Plato's time: the fourth: 4/3, the fifth 3/2 and the tone 9/8. To obtain a musically harmonious arrangement, all that had to be found was the octave 2/1 that fills the interval between the fourth (4/3) and the fifth (3/2) and the *leimma* (= in ancient Greek, "that which remains") 256/243 that fills the interval remaining between the two tones (9/8). Hence this table:

a	1	9/8	81/64	4/3	3/2	27/16	243/128	2
2a	2	9/4	81/32	8/3	3	27/8	243/64	4
4a	4	9/2	81/16	16/3	6	27/4	243/32	8
8a	8	9	81/8	32/3	12	27/2	243/16	16
16a	16	18	81/4	64/3	24	27		

$$9/8 \quad 9/8 \quad 256/243 \quad 9/8 \quad 9/8 \quad 9/8 \quad 256/243$$
$$4/3 \qquad\qquad 4/3$$
$$4/3 \qquad\qquad 3/2$$
$$2/1$$

Considered only from a musical point of view, the mathematical construction of the world soul would therefore comprise 4 octaves, a fifth and a tone:

$$2/1 \cdot 2/1 \cdot 2/1 \cdot 2/1 \cdot 3/2 \cdot 9/8 = 27$$

However, it should be observed that Plato's intention was not to produce a theory of the type of music that the heavenly bodies might emit.

Borrowing the idea of the "harmony of the spheres" perhaps from the Pythagoreans, Plato extrapolates. Knowing that musical harmony is governed by mathematical laws, he postulates that heavenly bodies, whose movements present the permanence and regularity that mathematical means provide in music, are also governed by such laws: the sort of mathematics that "works" so well in music, music which means harmony by antonomasy, should "work" just as well in astronomy.

The Role of the World Soul

The stakes are important, since the role of the world soul is to explain the how and why of the ordered movement of the sensible world. The more the world soul is governed by rigorous mathematical laws, the more the movements affecting the sublunar sensible world will be ordered. As the above-described proportions apply to a series of integers that represent the orbital radius of each of the seven planets surrounding the earth, these proportions affect not only the arrangement of the heavenly bodies, but also, and above all, the speed of their revolution, since Plato believes that the speed of the revolution of a body varies in function of the length of the radius of the circle it describes. This permanence and this regularity are transmitted, in various degrees, by the heavenly bodies to sensible things, as will be shown.

Let us go back to the working demiurge. After having cut the metal sheet described above, the demiurge crosses the two resulting bands making them coincide in their middle, as in the figure of the Greek letter *X*:

Figure 1.2

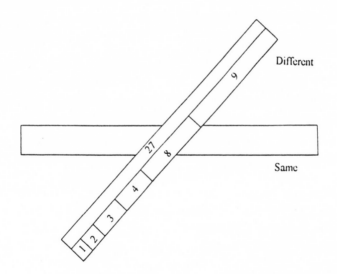

Next he bends these two bands and joins their extremities, thus forming two circles:

Figure 1.3

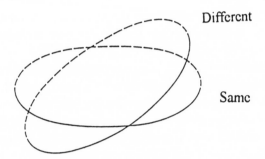

On the first circle, the circle of the "Same," move the fixed stars; the whole sphere of which the sensible world consists follows this movement, from east to west. On the circle of the "Different," move the seven "planets": Moon, Sun, Mercury, Venus, Mars, Jupiter, Saturn.[7]

Figure 1.4

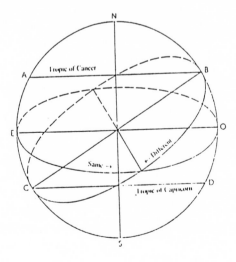

Comments

• AB is the Tropic of Cancer
• CD is the Tropic of Capricorn
• The movement of the "Same" is the movement of the world's sphere, going from left (= the east) to the right (= the west) on the equatorial plane (= EF).
• The Zodiac can be represented by a large band where the twelve signs are ordered, forming a ring that shares its center with Ecliptic (BC), and whose circumference follows the sphere's envelope.

The world body thus consists of an immense sphere that contains all sensible reality and beyond which there is nothing. This explains the perpetuity of the sensible world, for nothing can come from the outside to disturb or destroy it. Moreover, we must imagine that the fixed stars are in some way fastened to the internal surface of this sphere and moved with it; this explains why the movement of the fixed stars and that of the sphere that comprises the world body are identical.

After these manipulations, the demiurge moves on to the final operation, which consists in dividing the interior circle six times (*Timaeus* 36 d) in order to obtain seven unequal circles corresponding to the orbits of the following planets: Moon, Sun, Mercury, Venus, Mars, Jupiter, Saturn, the Earth remaining immobile at the center of the sensible world (*Timaeus* 40b-c).

Figure 1.5

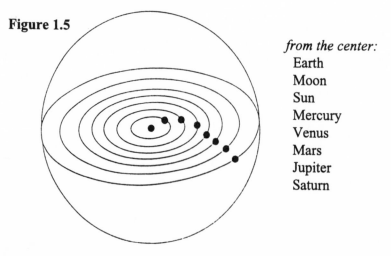

from the center:
Earth
Moon
Sun
Mercury
Venus
Mars
Jupiter
Saturn

The resulting construction is fitted by the demiurge into the sphere which encompasses the sensible world, making sure that their respective centers correspond perfectly (*Timaeus* 36d-e, cf. 34b).

By so doing, the demiurge endows the sensible world with a principle that accounts for all ordered movements, whether they be of a physical (astronomy) or a psychical nature (knowledge). Indeed, by means of the circle of the "Same," the world soul establishes direct contact with the world of intelligible forms, and, with the circle of the "Different," it is put in contact with the sensible world. Since it is a living thing endowed with reason, the sensible world can autonomously order its own movements (i.e., changes), in a "rational" way.

Thus, in the *Timaeus* (38c-39e), Plato endeavors to put forward a complete astronomical system based exclusively on circular movement, a system that remained viable until Kepler.[8]

Corollary: The Definition of Time

The movement of the heavenly bodies permits Plato to establish several standards for the measure of time. The movement of the circle of the Same brings about the alternation of day and night. The movement of the Moon produces the monthly sequence, and that of the Sun, the yearly sequence. But Plato is even more daring. He hypothesizes a temporal duration measured by the revolution of the five other planets, and a "great year" corresponding to the return of all the heavenly bodies to their initial position (*Timaeus* 39c-e).

Since in the world of Becoming, only images exist, time, which is the measure of Becoming must present two characteristics: it is 1) an image, but 2) an image ordered by numbers. As it is directly associated with the unceasing change of the sensible world, time becomes an object of knowledge only if a resemblance is found between it and an intelligible Form, a resemblance that can be expressed with the help of a mathematical relationship of the type:

eternity/unity = time/the diversity of numbers.

Hence this famous definition: "When he [the demiurge] ordered the heaven, he made, of eternity that abides in unity, an everlasting likeness moving according to number, that to which we have given the name 'time'." (*Timaeus* 37d) Time is therefore indistinguishable from the world soul, for it is engendered by the movements of all the circles of the world soul; it is also indistinguishable from the world body, for the standards of temporal measurement are provided by the revolutions of the heavenly bodies. It follows that the sensible world can not be engendered in time since it itself engenders time. Sensible world and time co-exist;[9] there can be no time before the sensible world.

Plato's Theory of Matter and the Cosmological Axioms[10]

An astronomical system is developed in the cosmology of the *Timaeus*. Plato's aim is to explain *everything*, including the underlying microscopic world and the complex world of human experience; hence the need to take into consideration not only physical phenomena but also biological and psychological phenomena. Consequently, a theory of matter must be advanced, implying new axioms.

Let us begin by recalling an above-mentioned axiom.

Axiom T9

> **Sensible particulars, including heavenly bodies, are made up of four elements only: fire, air, water and earth.**

Greek physics was based upon this axiom. Plato wants to describe the origin of these elements, and moreover their mathematical origin. In so doing, he is aware of being truly original:

> ... what I [= Timaeus] must now attempt to explain to you is the distinct formation of each [of the elements] and their origin. The account will be unfamiliar; but you [= Socrates, Hermocrates, Critias] are schooled in those branches of learning which my explanations require [i.e., mathematics], and so will follow me. (*Timaeus* 53b-c)

The astonishing modernity of this endeavor has not gone unnoticed. In an article published in 1955, Werner Heisenberg compares modern particle physics with the theory of elementary material components laid out in the *Timaeus*:

In what follows, the particular form of the philosophy of nature propounded by Plato in the *Timaeus* shall be considered, with special emphasis on a characteristic trait of this approach which resurfaces in modern atomic physics, in the theory of elementary particles, and which there plays an important role. The mathematical forms by which we nowadays represent the elementary particles ... are more complex than the geometric forms postulated by the Greeks. But essentially, in both cases these forms originate in certain simple mathematical basic requisites. And one must not forget that on this point the program of modern physics has yet to be completed. The similarity between Plato's ideas and modern atomic physics appears furthermore in a different context. Should one inquire in Plato's [philosophy], what is the content of his [geometric] forms, out of what stuff they are ultimately made, one gets this answer: out of mathematics.

Heisenberg continues:

In the final analysis, in both cases, the notion of matter is essentially a mathematical concept. The most fundamental kernel of all that is material is for us, as well as for Plato, a [mathematical] form, and not some material content.[11]

Heisenberg clearly saw the amazing similarity between Plato's theory of matter and its contemporary counterpart, and this on a fundamental axiomatic level. Indeed, the theory of matter in the *Timaeus* is based on the following axioms.

Axiom T13

> **The entire universe can be reduced to discrete, elementary components.**

Axiom T14

> **Such elementary components are infinitely small and there-fore "invisible."**

Axiom T15

> **Fundamentally, these components are mathematical entities.**

Axiom T16

> **These distinct elementary components are few in number, simple, indiscernible[12] and indestructible.**

Axiom T17

> **These entities are the ultimate components of all sensible particulars; everything in the universe is made from them.**

Axiom T18

> **In the universe, all observable phenomena, all that Plato calls "change" can be reduced to interactions between the elementary components.**

Axiom T19

> **These interactions can be described exclusively in mathematical terms.**

Axiom T20

No complexity factor exists associated to a given dimension-scale. For each level, from the dimension-scale of microscopic realities invisible to the naked eye, to that of the gigantic objects of astronomy, the explanation of observable phenomena is only to be found in the elementary components obeying the same mathematical laws.

Corollary to Axiom T20

The same mathematical laws and the same elementary components constitute the only substratum to which even hyper-complex phenomena, the kind detectable at the human scale, can be reduced.

Once this is accepted, the set of phenomena described by biology, by physiology, by pathology and even by psychology, can be reduced to these components.

As was the case with a large part of the fundamental ideas put forward by Plato in the *Timaeus*, this theory of matter was neither accepted nor understood over the following centuries. These ideas were rediscovered only in the nineteenth century, leading to the model of matter advanced by modern physics.[13] Plato's presuppositions reduce all sensible objects to "appearances" resulting from the combination of elementary particles, few in number, simple, and the mathematical rules that govern their correlations. Aristotle never accepted such a radical reduction. For him, in the sensible world we find not only particulars, but also individuals, each one characterized by its own indivisible form.

Remaining close to tradition, Plato admits that the universe is constituted by four elements exclusively: fire, air, water, and earth (axiom T9). Having arrived at this point, he is compelled to secure the consistency of axiom T9 with axioms T13 through 20. This can solely be achieved by reducing the four traditional elements to purely mathematical entities, and by formulating mathematically the inter-relationships prevailing amongst them.

According to axiom T10, if fundamental elementary components of the universe do exist, these components must be as perfect as possible. This amounts to say that they can be nothing else than mathematical objects (axiom T15).

> Now we must think of all these bodies [the elements] as so small [axiom T14] that a single body of any one of these kinds is invisible to us because of its smallness: though when a number are aggregated the masses of them can be seen [axiom T13]. And with regard to their numbers, their motions, and their powers in general, we must suppose that the god adjusted them in due proportion, when he had brought them in every detail to the most exact perfection permitted by *anagke* willingly complying with persuasion (*Timaeus* 56b-c).

Plato's solution consists in having the four traditional elements put in correspondence with four of the five regular polyhedra known at his time, these polyhedra being considered the most perfect geometrical objects, in accordance with the desire of the good demiurge to make a world out of elements bearing the greatest resemblance to the intelligible forms (axiom T10). The symmetry of these polyhedra should be noted; indeed they remain unchanged under rotation about several axes.

As the next step, Plato shows how the faces of these polyhedra, equilateral triangles for the tetrahedron, the octahedron and the icosahedron and square for the cube, can be ultimately constructed using only two kinds of right-angled triangles, the isosceles and the scalene.

Figure 1.6

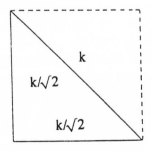
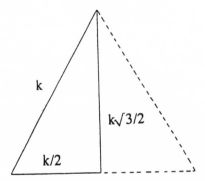

These two fundamental elementary right-angled triangles are used by the demiurge to make up the faces of the polyhedra, the square and the equilateral triangle.

Here is how a square results from the combination of four right-angled isosceles triangles: "This body is formed by four isosceles; the sides of their right angles join again in a center and form a quadrangular, equilateral figure." (*Timaeus* 55b)

Figure 1.7

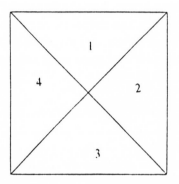

The construction of the other surface, an equilateral triangle, is more complicated:

Figure 1.8

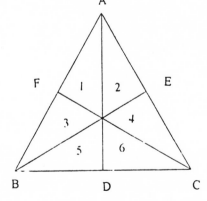

It has never been explained why Timaeus needed six right-angled isosceles triangles to make an equilateral triangle when two would have sufficed.

From the equilateral triangles, the demiurge constructs three of the four regular polyhedra: the tetrahedron, the octahedron and the icosahedron, associated respectively with fire, air and water: and from the squares, he constructs the cube associated with earth.

Regarding the construction of the tetrahedron, Timaeus explains: "If four equilateral triangles are put together, their plane angles meeting in a group of three make a solid angle, namely the one (180°) that comes next after the most obtuse of plane angles. When four such angles are produced, the simplest solid figure is formed, whose property is to divide the whole circumference into equal and similar parts." (*Timaeus* 54e-55a)

Figure 1.9

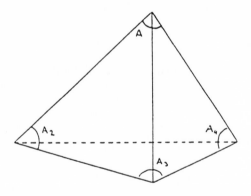

Timaeus continues, describing the construction of the octahedron: "A second body is composed of the same triangles. When they are combined in a set of eight equilateral triangles and yield a solid angle formed by four plane angles the second body is complete." (*Timaeus* 55a)

Figure 1.10

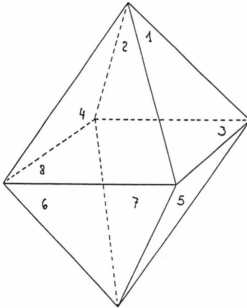

The description of the icosahedron follows: "The third body is composed of one hundred and twenty of the elementary triangles fitted together, and of twelve solid angles, each contained by five equilateral triangular planes; and it has twenty faces which are equilateral triangles." (*Timaeus* 55a-b)

Figure 1.11

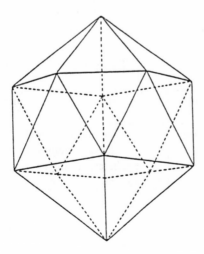

Lastly, Timaeus describes the cube: "Six quadrangles, joined together, produced eight solid angles, each composed by a set of three plane right angles. The shape of the resulting body was cubical, having six quadrangular equilateral planes as its faces." (*Timaeus* 55c)

Figure 1.12

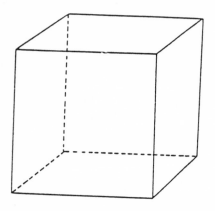

And he concludes with an evocation of the dodecahedron, the regular polyhedron that is most closely related to the sphere: "There still remained one construction, the fifth; and the god used it for the whole, making a pattern of animal figures thereon." (*Timaeus* 55c) Consequently none of its properties is mentioned, only a fleeting allusion that we will not develop here.

   Here is a table summarizing all the properties of the four polyhedra with which the four elements are associated.

Table 1.2

Element	regular solid	number of faces	number of triangles*
fire	tetrahedron	4 equilateral triangles	24 scalenes
air	octahedron	8 equilateral triangles	48 scalenes
water	icosahedron	20 equilateral triangles	120 scalenes
earth	cube	6 squares	24 isosceles

* right-angled triangles

The value of "k," (cf. Figure 1.6) taken as a unit of measure of the hypotenuse of each of the two fundamental right-angled triangles, remains undetermined (cf Axiom T21: The universe is not uniform).

Table 1.2 displays the correspondence between the elements, the polyhedra and their fundamentally constitutive surfaces. Such correspondences allow the formulation of laws that explain how these elements are transmuted one into another. This alchemy, which allows us to understand the generation and the corruption occurring in the sensible world (axioms T15 through T17) is based on this presupposition: the two types of elementary right-angled triangles can neither be created nor destroyed (axiom T16). Consequently, in any transformation, the number of triangles of each sort involved in the reaction is conserved. Moreover, only elements corresponding to polyhedra formed from the same sort of right-angled triangle can be transmuted one into another. It follows that water, air and fire can be transmuted one into another, but not into earth which can only undergo processes of decomposition and recomposition, an inference Aristotle violently denounced.[14]

Table 1.3 describes the fundamental laws of "transmutations," where the symbol ▲ stands for the basic equilateral triangle.

Table 1.3

1 [fire] = 4 ▲
2 [fire] = 2 x 4 ▲ = 8▲ = 1 [air]
1 [fire] + 2 [air] = 4 ▲ + 2 x 8▲ = 20▲ = 1 [water]
2 ½ [air] = 2 ½ x 8▲ = 20▲ = 1[water]

It is interesting to realize how this solution, expressed in axiom T19, was imposed upon Plato by the limits of the known mathematics of his time. These limits, as Plato was aware, appear clearly when considering the problem of the extraction of square or cube roots. The following analysis is based on this passage:

Now that which comes to be must be bodily, and so visible and tangible; and nothing can be visible without fire, or tangible without something solid, and nothing is solid without earth. Hence the demiurge, when he began to put together the body of the universe, set about making it of fire and earth. But two things alone cannot be satisfactorily united without a third; for there must be some bond between them drawing them together. And of all bonds the best is that which makes itself and the terms it connects a unity in the fullest sense; and it is of the nature of a continued geometrical proportion to effect this most perfectly. For whenever, of three numbers, the middle one between any two that are integer or power of an integer is such that, as the first is to it, so is it to the last, and conversely as the last is to the middle, so is the middle to the first, then since the middle becomes first and last, and again the last and first become middle, in that way all will necessarily come to play the same part towards one another, and by so doing they will all make a unity. Now if it had been required that the body of the universe should be a plane surface with no depth, a single mean would have been enough to connect its companions and itself; but in fact the world was to be solid in form, and solids are always conjoined, not by one mean, but by two. Accordingly the demiurge set water and air between fire and earth, and made them, so far as was possible, proportional to one another, so that as fire is to air, so is air to water, and as air is to water, so is water to earth, and thus he bound together the frame of a world visible and tangible. (*Timaeus* 31b-32b)

In this difficult passage, two types of proportions are discussed: one involved in a two-dimensional world and one involved in a three-dimensional world.

Plato deals first with a two-dimensional world, which raises the problem of its area. The determination of the area of simple figures such as the square had led the Greek geometers to the problem of the irrationals. They found that the height h of a right-angled triangle is the proportional mean between the segments x and y that this height determines on the hypotenuse. As $h^2 = x \cdot y$, $h = \sqrt{x \cdot y}$, it follows that the three terms of this equation are seldom all rational numbers. This difficulty was, in Plato's time, linked to the problem of the duplication of the square, as can be seen particularly in the *Meno* (81a-84b), where it is proved that to double the area of the square with sides equal 1, a segment of length $\sqrt{2}$ was required. Square roots in Plato's time could be extracted, but only for numbers smaller that 17 or 19. In a three-dimensional world however, new difficulties appear. In the quoted passage, Plato recalls the problem of the duplication of the volume of a cube, much discussed in his time. Indeed, Hippocrates of Chios (470-400 BC) had shown that the duplication of the volume of a cube requires the previous determination of the two proportional means that can be inserted between any two numbers. But these two proportional means, "necessary for harmonizing these two volumes," as he puts it, are also cube roots and thus most often irrational quantities. And, as the problem of geometrically constructing such quantities had not been resolved in Plato's time, we understand why several lines further on he writes: " ... the demiurge set water and air between fire and earth, and made them, so far as was possible, proportional to one another." (*Timaeus* 32b).

This avowal of relative impotence is understandable if one looks at the relationships between the volume V and the surface S of the four regular polyhedra (whose side is a) of Plato's world, as this table shows:

Table 1.4

Polyhedra	V	S
Tetrahedron	$1/12\ a^3\ \sqrt{2} = 0.1178a^3$	$a^2\sqrt{3}$
Hexahedron	a^3	$6a^2$
Octahedron	$1/3\ a^3\sqrt{2} = 0.4714a^3$	$2a^2\sqrt{3}$
Dodecahedron	$1/4\ a^3\ (15+7\sqrt{5}) =$ $7.6631a^3$	$3\sqrt{25+10\sqrt{5}}\ a^2$
Icosahedron	$5/12\ a^3\ (3+\sqrt{5}) =$ $2.1817a^3$	$5a^2\sqrt{3}$

In any case, neither Plato nor any of his contemporaries would have grasped a single line of Table 1.4, because for them extraction of the cube root was impossible, and the very laborious extraction of the square root could only go up to 17 or 19.[15] The limits of Plato's cosmological system are the limits of the mathematics he knew.

In summary, up to this point, Plato has set forth in detail his overall cosmological theory: a benevolent demiurge, making use of the most symmetrical elementary components, secures a verisimilar account of the world by persuading *anagke* that these components shall be ruled by mathematical proportions as far as possible.

The most complex substances result from a combination of these elementary components (axiom T15), and their properties can be deduced from their geometrical and mathematical characteristics. Timaeus explains that earth (corresponding polyhedron, cube) is "the most stable" element (*Timaeus* 55e).

> ... and of the remainder the least mobile is water, the most mobile is fire, and the intermediate figure is air. Again, we shall assign the smallest polyhedron to fire, the largest to water, and the intermediate to air; and again the polyhedron with the sharpest angles is fire, the next to air, the third to water. Now taking all these figures, the one with the fewest faces (tetrahedron) must be the most mobile, since it has the sharpest cutting edges and the sharpest point in every direction, and moreover the lightest, as being composed of the smallest number of similar parts; the second (octahedron) must stand second in these respects, the third (icosahedron), third. Hence, in accordance with genuine reasoning as well as probability, among the solid figures we have constructed, we may take the tetrahedron as the element or seed of fire; the second in order of generation (octahedron) as that of air; the third (icosahedron) as that of water. (*Timaeus* 56a-b)

Timaeus relates the geometrical properties, possessed by the regular polyhedra to which the four elements correspond, to the physical characteristics that these elements present when they are perceived by the senses. But does this theory work? The entire last two thirds of the *Timaeus* represents Plato's strenuous attempt to prove that it does.

To begin with, the following series of the properties of certain sensible particulars are explained by Plato in terms of the four elements exclusively. The fundamental idea is to establish a relation between the form, the size and the movement of the polyhedra and the sense-perception to which they give rise in sense-organs presumed to be constituted by the same elements. Thus, Plato admits that sense-perception can directly generate a measure, and, on this assumption, he proceeds to test his theory.

Varieties of the four elements	**Sensible qualities as they appear to sense-perception**
Varieties of fire (= tetrahedron)	
flame	hot
light	vision
glow	?
Varieties of air (= octahedron)	
aether	bright
mist	odors
darkness	black
without a specific name	sounds
Varieties of water (= icosahedron)	
i) when water is a liquid	
juices	taste
many without a name	
some with a name	
wine	heat
oil	bright, smooth
honey	sweet
opos	sour & pungent
ii) in an intermediate state	
when melting	
water	soft
when freezing	
hail	cold
ice	cold
snow	cold
frost	cold

iii) as a fusible solid
    gold   yellow, heavy, soft
    adamas   black, hard
    copper   bright, light, hard
    etc.

Varieties of earth (= cube)
  i) when strained through water
      stone   heavy
      diamond?   transparent
  ii) when deprived of moisture by air
    earth not soluble in water
      earthenware
      lava?   dark
    earth soluble in water
      soda   taste
      salt   taste
  iii) bodies composed of earth and water
    glass
    wax
    incense

Plato must admit that sense-perception can *directly* generate a measure, since he has no theory of measurement. As a matter of fact, in Plato's time, no universal standard units of measure existed, by means of which alone a mathematical explanation of physical situations is possible. Therefore Plato had to resort to this artifice: since everything in the world, the sense-organs included, is constituted solely by mathematical building blocks, a mathematical interpretation of sense-perception is, in principle, feasible.

At this stage of the construction of the model, a last group of axioms is required in order to complete Plato's cosmology.

The Cosmological Axioms

Axiom T21

The universe is not uniform.

The movement observed in the universe, its perpetual change, is a consequence of the non-uniform size of the elementary polyhedra.

> ... motion will never exist in a state of homogeneity or uniformity. (*Timaeus* 57e)

Since the fundamental magnitude "k" (cf. Figure 1.6), the size of the hypotenuse, of the elementary triangles remains undetermined, it follows that the sizes of the elementary polyhedra that make up all sensible things must also be different. This lack of uniformity constitutes the cause of the change that always affects the sensible world, change that the world soul attempts to order, but only there where it rules as its mistress and governor.

Axiom T22

> **In the sensible world, there is no vacuum. (*Timaeus* 58a, cf. 79c).**

Axiom T23

> **The sphere of the world encloses all that is corporeal. The four elements are spread out, within this sphere, in four concentric layers (*Timaeus* 33b, 53a, 48a-b).**

These four concentric layers are pulled by the circular movement which animates the entire sphere. As there is no vacuum (axiom T22), the elementary polyhedra cannot spread out beyond the sphere. And, in its interior, they can only move within the always-filled interstices originating from the absence of homogeneity of the polyhedra (axiom T21). Hence a chain reaction, resulting from the condensing process, thrusts the small bodies together into the interstices between the large ones (*Timaeus* 58b, cf. 76c and *Laws* X 849c). This chain reaction produces the following process:

> Accordingly, when the small [polyhedra] are set alongside the large, and the lesser disintegrate the larger, while the larger

cause the lesser to combine, all are changing the direction of their movement this way and that, towards their own regions; for each, in changing its size, changes also the situation of its region. In this way, then, and by these means there is a perpetual safeguard for the occurrence of that heterogeneity which provides that the perpetual motion of these bodies is and shall be without cessation. (*Timaeus* 58b-c)

This double movement governs every mathematical transformation of one polyhedron into another. In Plato's physics such transformations between polyhedra can take only two well-defined forms: division/condensation and combination/disintegration (cf. Table 1.3 above).

The sensible world is governed by a soul presenting a particularly rigorous mathematical structure; what is more, the demiurge fashioned the *khora* mathematically, shaping the four elements as regular polyhedra. Hence, transformations of one polyhedron into another can be reduced to mathematical interactions and correlations. Change, which characterizes sensible things as opposed to intelligible Forms, acquires, thanks to mathematics, certain predicates of the intelligible world in which it partakes; the sensible world is thus cloaked in that permanence and regularity that distinguishes the intelligible world. In such a way, mathematics accounts for the participation of sensible things in intelligible Forms.

Plato's Explanation of the Complexity of Sensible Things

The mathematization of sensible particulars extends as well to areas presenting the greatest complexity, in particular the human body, of which Plato advances a purely mathematical description. We will only discuss the composition of man's body by the demiurge; but neither the problem of its functioning nor that of its destruction by disease will be dealt with[16] here.

Like all other sensible things, the human body is made up of essentially two types of right-angled triangles forming the elementary surfaces of the four regular polyhedra corresponding to the four elements. For the generation of the two fundamental components of the human body, the marrow and the flesh, the demiurge uses right-angled triangles of two different qualities.

On one hand, marrow, a substance in which the soul comes to anchor itself and which is made up of "unwarped and smooth triangles."[17] On the other hand, flesh, a substance that corresponds to the corruptible, is made up of ordinary triangles.

To make the marrow, the demiurge chooses regular and smooth triangles able to produce fire, water, air and earth possessing the most exact form. Mixing together these triangles, he composes the marrow with which he makes the brain, the spinal chord, and bone marrow. Then, watering and thinning out pure earth, passed through a screen, and then mixing this earth with the marrow, he makes the bony substance which he uses to fashion the skull, the spinal column and all the other bones.

Next, using elements composed of ordinary triangles, the demiurge produces the flesh by mixing water, fire, and earth, to which he adds a leaven formed from salt and acid, themselves made from ordinary triangles. When the flesh dries, a film appears, the skin. Moisture, escaping from the holes pierced in the skin by fire and forced under the skin by air, takes root on the skull and produce hair. With a mixture of bone and flesh without leaven, the demiurge makes the sinews which he uses to attach the bones together. And, lastly, with a mixture of sinews, skin and air, he makes the nails.

Man's body is therefore reduced to the regular polyhedra, themselves composed of surfaces resulting from grouping two types of right-angled triangles: isosceles and scalene. The mathematical properties of these two elementary triangles "explain" the difference between the marrow, anchoring point of the soul's immortality, and the flesh, a mortal substance. Moreover, the destruction of the human body by diseases and the (psychological) problems raised in the soul by its union with a corruptible body are similarly explained in "mathematical" terms.

This description of the human body, which may seem so strange to us, appears, in the final analysis, nothing else than an extreme application, to the microcosm, of the above-mentioned axioms.

Experimental verification in Plato's time

In the preceding sections, 23 axioms underpinning the cosmological model advanced by Plato in the *Timaeus* have been formulated; they are the basic postulates of Plato's mathematical model of the universe. To confirm his theory, Plato must establish the connection between the propositions deduced from these axioms and observational data. His theory advances a model of how the heavenly bodies move and of their relative distances and velocities. It has also been shown that his axioms allow him to establish the laws by which division and condensation, combination and disintegration occur in the physical world. Furthermore, he "deduces" from his axioms the qualities of a large number of substances such as ice, copper, wax, etc. And finally, he "deduces" a large number of properties of the human body. What is left to be done is to set up an experimental protocol suitable for the verification of this theory. But Plato only offers direct sense-perception. We thus ask: could he have done better? What was the status of experimentation in Plato's time? What technical means were available for such a purpose?

The Pythagoreans, whom Plato probably encountered on his first voyage to southern Italy and Sicily, had apparently succeeded in showing that a mathematical relationship exists between the lengths of the strings of a consonant lyre. These results were incorporated in Plato's description of the world soul. Furthermore, it seems certain that Eudoxus, one of Plato's disciples at the Academy, had managed to construct a model that described the "retrograde" movement of the planets with the help of two concentric spheres executing their rotation in opposite directions around two inclined axes; he is thus considered the father of mathematical astronomy. Apart from these elementary results, no other example of what we moderns refer to as "experimental verification" is reported.

In ancient Greece, experimental verification is almost never brought into play to confirm a mathematically formulated theory. Several reasons explain this situation. First, the scarcity of appropriate abstract units of measure, and the absence of corresponding instruments of measure. Another no less decisive factor has already been mentioned:

mathematics in Plato's time was in a particularly primitive state, and several of its essential developments were still lacking.

Standards and instruments of measure in Plato's time

The measuring operation is science's most fundamental practice. To progress, science must, at the outset, define its units of measure. The discovery of such units of measure for temperature, acceleration, energy, electrical charge, entropy, information (measured in *bits*), etc., and the perfecting of instruments enabling their measure, represented each time a decisive step for the development of science.

However, in Plato's epoch, the only known standards of measure were those related to length, weight, volume and time.

Measures of length

Revealing their degree of primitiveness, the units of length were directly associated with parts of the human body: *daktulos* (the finger), *palaiste* (the palm). The *pous* (foot) was the basic unit of measure for length, which however could vary, for, depending on the city, feet of 0.296 m., 0.328 m. and 0.32045 m. were used. The *pekhus* (cubit) = 1.5 feet and the *bema* (pace) = 2.5 feet are further examples. Longer length-measures were in use, such as the *stadium* = 600 feet, the Olympic measure, which was equivalent to 192.27 m. Locally, a certain standardization was sought, as archaeologists have observed when carrying out measurements on the ruins of buildings from that time. But no reference to a "standard of length" has been found.

Measures of weight

Greek measures of weight are directly associated with commerce in general, and with the equally necessary weighing of precious metal coins. This is why the bronze weights were kept in the temple of Athena on the Acropolis. Ten magistrates known as *metronomoi* had the task of ensuring that these unities of weight were respected: their names can sometimes be found inscribed on the standard weights. Let us mention as units of weight: the *khalkos* = 0.09 grams, the *obolos* = 8 *khalkoi* =

0.72 grams; the *drakhme* = 48 *khalkoi* = 4.32 grams; the *mna* = 100 *drakhmai* = 432 grams; and the *talanton* = 60 *mnai* = 25.92 kilograms. The measuring operations were carried out on balances with a central pivot, a beam and two plates: in ancient Greek, the name of this less than accurate instrument is "*talanton.*"

Measures of volume

Measures of volume, conceived exclusively for the exchange of merchandise, were even less accurate since they depended entirely on several external factors relating to the product in question. For example, 1 *medimnos* = 52.53 liters of dry matter, but 1 *metretes* = 39.39 liters of liquid matter.

Measures of time

Nothing resembling an accurate time-measuring instrument appears in European history before the seventeenth century, prior to the invention and utilization of the pendulum. The isochronism of the pendulum was discovered by Galileo and developed by Christian Huyghens. In Plato's epoch, there existed but two time measuring methods: sun dials, examples of which can already be found in Egypt about 2000 BC, and water clocks or clepsydras, already used in Egypt about 1600 BC. These two instruments were particularly inadequate for experimentation: the first could only be used outdoors and by daylight; and the second provided different results according to a series of factors: the viscosity of the liquid used, the ambient temperature, the diameter of the opening through which the liquid flows, etc.

On the other hand, the determination of the length of the solar year in terms of lunar months presented a daunting problem, which was resolved only under pope Gregory XIII in 1582. In Plato's time, months counted 29 or 30 days, the first day coinciding with each new moon. Consequently, the year was 11.25 days too short, a situation that meant the establishing of yearly cycles of 12 and 13 months respectively. In 423 BC, Konon established a cycle of 19 years for a total of 235 months. To further complicate things, years were named after the *eponymous archon*, the highest ranking magistrate in Athens. As a result, to refer to a date only several years previous, a long explanation was required, especially if the Olympiad, that is, the four-year interval separating the Olympic Games, was taken into account.

The numbering system in Plato's time

In written texts, the ancient Greeks used as numbers the 24 letters of their alphabet, to which they appended three other symbols for a total of 3 x 9 = 27 symbols, allowing them to write numbers up to 999. To attain higher numbers, sub-indices were attached to these symbols. This enabled the Greeks to reach 900,000. For practical reasons, this system could generally not exceed 100,000. For higher quantities, the adjectives *murioi*, ten thousand, and the adverbs *dekakis*, ten times, and *eikesakis*, twenty times, etc. were used. In ancient inscriptions another way of indicating numbers appears, showing similarities to the Latin numbering system. For example, the number 1957 was written:

Figure 1.13

ΧΓΗΗΗΗΗΓΙΙ

Finally, the Greeks were not aware of the zero.

Under these circumstances, it should be obvious that, even if they did reach a fairly advanced level of geometry, even if they did succeed in accomplishing technical feats, as witnessed in their architecture, and their sculpture, and even if their navigational methods did require the use of technical procedures, primitive though they were, the Greeks of Plato's epoch did not have access to the tools that would have enabled them to define and implement experiments capable of verifying their hypotheses in the realm of physics. This remark notwithstanding, their arithmetical prowess must have been superb: to calculate the square root of some number using only letters and knowing nothing about the zero is quite an achievement.

Experimentation in the *Timaeus*

In spite of the above arguments, two comments are warranted regarding the possibility of experimental verification in the *Timaeus*.

First, it is evident that Plato never thought of submitting any particular aspect of his theory to a real test, that is, within the framework of a purely local experiment, controlled and repeatable. Experiment today presents the following characteristics: over the course of an experiment, a very limited number of parameters are varied while supposing that the rest of the universe, with its enormous complexity and high degree of freedom will not exert an influence on the experiment under way: *ceteris paribus*, "all the rest doesn't count." To arrive at this *ceteris paribus*, all the ingenuity of the experimenter must be brought to bear, which sometimes leads him to construct gigantic and very complex instruments. This absolutely decisive procedure of questioning nature eluded Plato who, when he refers to a physical phenomenon, states:

> But any attempt to put these matters to a practical test (*skopoumenos basanon lambanoi*) would argue ignorance of the difference between human nature and divine, namely that divinity has knowledge and power sufficient to do so ... but no man is now, or ever will be, equal to either task.   (*Timaeus* 68d-e)

For Plato, experimental verification therefore implies an *exact* reproduction of nature, just as impossible a task for us as for him. The history of science would have been radically different if Plato had used an experimental method enabling him to limit his research to only one element of reality at a time. It is interesting to note that the expression Plato employs to designate "experiment," is *basanon lambanein,* an expression designating the "question," the torture to which a slave was submitted in order to make him confess a crime of which he was suspected.

Secondly, a certain number of propositions, appearing in the *Timaeus*, which are based on observation and which are in principle testable, can be quoted:

> Of all these fusible varieties of water [= metals, because they melt], as we have called them, one that is very dense, being formed of very fine and uniform particles, unique in its kind, tinged with shining and yellow hue, is gold ... . Another has particles nearly like those of gold, but of more than one grade; in point of density in one way it surpasses gold and it is harder because it contains a small portion of fine earth: but it is lighter by reason of containing large interstices. This formation is copper, one of the bright and solid kinds of water. The portion of earth mixed with it appears by itself on the surface when the two substances begin to be separated again by the action of time; it is called verdigris. (*Timaeus* 59b-c)

> Sound we may define in general terms as the stroke inflicted by air on the brain and blood through the ears and passed on to the soul. ... A rapid motion produces a high-pitched sound; the slower the motion, the lower the pitch. If the motion is regular, the sound is uniform and smooth; if irregular, the sound is harsh. According as the movement is on a large or a small scale, the sound is loud or soft. Consonance of sounds must be reserved for a later part of our discourse. (*Timaeus* 67b-c)

These examples indicate that, despite the many deficiencies on the technical level and lack of adequate instruments of measure, Plato, in the *Timaeus*, sets forth propositions that conflict neither with logic nor with sensible experience. Thus, even if observation and verification, as understood by moderns, were not technically possible in Plato's time, they were not systematically rejected.

With these remarks, we conclude the presentation of the axiomatic system on which the cosmological model presented by Plato in the *Timaeus* is based. In the following section, the most developed cosmological model, the standard Big Bang model, will be discussed from a perspective as similar as possible to the one developed here.

NOTES

1. We prefer to translate *eidos* and *idea* by "intelligible Forms," for, since Descartes, the English term "idea" evokes, in philosophy, a representation. For Plato, *eidos* and *idea* designate a reality, the true reality.

2. Here, we agree with F.M. Cornford: "Whereas you can always ask the reason for a thing's existence and the answer will be that it exists for the sake of its goodness, you cannot ask for a reason for goodness; the Good is an end in itself." (*Plato and Parmenides*, London 1939, p. 132).

3. G.E.R. Lloyd, *Magic, reason and experience. Studies in the origin and development of Greek science*, Cambridge 1979.

4. The fact that the Platonic demiurge is not omnipotent bothered the Christians who, under bishop Tempier in Paris in the thirteenth Century, condemned these ideas as heretical. Let us consider, as an example, what article 93 condemns as heretical: Quod aliqua possunt casualiter evenire respectu causae primae; et quod falsum est omnia esse praeordinata a causa prima, quia tunc evenirent de necessitate. 1. This article asserts the existence of chance, even with regard to God, and rejects the *praeordinatio* of events by the primary Cause, for this praeordinatio would entail their necessity. 2. Divine Providence, which governs the entire order of natural causes, implies that nothing eludes his *praeordinatio*; consequently, nothing is fortuitous with regard to God. Chance is found only at the level of secondary causes. Cf. *Enquête sur les 219 articles condamnés à Paris le 7 mars 1277*, text and commentary by Roland Hissette, Louvain/Paris 1977.

5. The Greek word *kalon* means also "desirable" and "beautiful" and can here be so translated.

6. In a sense, all the mathematical formulae we have included represent anachronisms, especially because they refer to notation unknown in Plato's time.

7. The modern names became usual at a much later date.

8. The *Timaeus* attempts to represent the movements of the heavenly bodies exclusively as a combination of circular movements (for a list of all the references cf. Luc Brisson, *Le même et l'autre dans la structure ontologique du* Timée *de Platon*, Paris 1974, p. 394-395); this is a project which has dominated astronomical research for the twenty following centuries. More than ten centuries after Plato, Ptolemy could

still proclaim: "Now that we are about to demonstrate in the case of the five planets, as in the case of the Sun and the Moon, that all of their phenomenal irregularities result from regular and circular motion - for such befit the nature of divine beings, while disorder and anomaly are alien to their nature - it is proper that we should regard this achievement as a great feat and as the fulfillment of the philosophically grounded mathematical theory [of the heaven]" (Ptolemy, *Syntaxis mathematica* 9.2, in *Opera omnia*, J.L. Heiberg ed., vol. I, par. 2, p. 208.) Quoted by G. Vlastos, *Plato's Universe*, Oxford 1975, p. 65, n. 107.

9. In the *Timaeus*, one must distinguish between the "narrative" that recounts the ordering of the sensible world by the demiurge, and the cosmological system based on the axioms we have enumerated. The appearance of time and ordering of the sensible world, should not be interpreted temporally, as Aristotle did in his critique *(De caelo* I 3; III 1). Even if a narrative implies an internal temporality in order to deploy itself, the *Timaeus* limits itself to describing the ontological dependency of time with regard to eternity. Briefly put, the origin of time is contemporaneous with that of the world, which, owing to this fact, cannot have a temporal origin.

10. To speak of "matter" in Plato is a controversial task. The first Greek philosopher to have developed a notion corresponding to what the Latin language authors called *materia* is Aristotle, who uses, to designate this notion, the term *hule*, which in ancient Greek designates, in its first meaning, "wood," and more specifically, "wood for construction." In the Platonic corpus, only the first meaning of this term can be found. In the *Timaeus*, the term closest to *hule* is *khora*, which we translate as "spatial medium," because the *khora* presents a spatial aspect, to the extent that it is that "in which" sensible things are found, and a constitutive aspect, to the extent that it is that out "of which" sensible things are made. In this perspective, matter is not exclusively the *khora*, but a *khora* organized either by chance, where *anagke* rules unhindered, or rationally, following the demiurge's intervention. Cf. Luc Brisson, *Le même et l'autre dans la structure ontologique du* Timee de *Platon*, Paris 1974, chapter 3.

11. "Im folgenden soll von der speziellen Form der Naturphilosophie die Rede sein, die in Platons Dialog *Timaios* niedergelegt ist, und es soll ein besonderer Zug dieser Vorstellungswelt erörtert werden, der im modernen Gebiet der Atomphysik, in der Lehre von den Elementarteilchen, wieder aufgetaucht ist und Bedeutung gewonnen hat. Die mathematischen Formen, durch die wir heute die Elementarteilchen darzustellen pflegen, sind (...) komplizierter als jene geometrischen Formen der Griechen. Grundsätztlich handelt es sich aber in beiden Fällen um Formen, die als Folge gewisser einfacher mathematischer Grundforderungen entstanden sind. Dabei is daran zu erinnern, dass das Programm der modernen Physik an dieser Stelle noch nicht zu Ende geführt ist. Die Aehnlichkeit zwischen den Vorstellungen bei Platon

und der modernen Atomphysik tritt auch noch an einer anderen Stelle in Erscheinung. Fragt man bei Platon, welches der Inhalt seiner Formen sei, aus welchem Stoff also seine regulären Körper schliesslich gemacht sein, so erhält man die Antwort: aus Mathematik." Heisenberg continues: "Letzen Endes wird also der Materiebegriff in beiden Fällen auf Mathematik zurückgeführt. Der innerste Kern alles Stofflichen ist für uns wie für Plato eine Form, nicht irgendein materieller Inhalt." (W. Heisenberg, "Platons Vorstellungen von den kleinstein Bausteinen der Materie und die Elementarteilchen der modernen Physik," *Im Umkreis der Kunst. Eine Festschrift für Emil Preetorius*, Wiesbaden 1953, p. 137-140).

12. All electrons are alike, as are protons, etc. Similarly, all the tetrahedra of fire, or all the cubes of earth etc. are indiscernible. Electrons differ by their energy, cubes, by the magnitude ("k") of their fundamental component triangles, etc.

13. But it should not be interpreted as 1 [fire] + 2 [air] corresponds to 1 [oxygen] + 2 [hydrogen] = H_2O [water]. Similarity is the result of pure coincidence.

14. *De caelo* III 7, 306 a1-7.

15. Table 1.4 shows that Table 1.3 is false. For example, according to the second line of Table 1.3, 2 particles of fire = 1 particle of air, while according to Table 1.4, 2 $(0.1178a^3) = 0.2356a^3$ and not $0.4714a^3$.

16. Luc Brisson, *Le même et l'autre dans la structure ontologique du* Timée *de Platon*, Paris 1974.

17. The Greek term *astrabe,* whose positive gave "strabismus" in English, refers to a triangle which is "not bent;" the Greek term *leia* refers to something which is "smooth" or "polished." Interpreters have not been able to explain the exact meaning of these terms within this context.

Part II

Contemporary Big Bang Cosmology

Is there such a thing as a singular, all-encompassing unique universe? And if so, what is its structure? is its size finite or infinite? did it have a beginning? is its duration eternal? and what is our place in it and what can we know about it? Such questions may be ranked among the most ancient of mankind's intellectual inquiries, and numerous answers have been proposed. But only in very recent times have these answers departed from a simple mythical narrative.

What makes modern cosmology a scientific inquiry is its *mathematical description of the universe*, which only became possible after achievements of Kepler and of Galileo during the Renaissance, a period coincidental with the rediscovery and translation of the complete Platonic writings. Modern cosmology can be defined as the search for a mathematical formulation representing the phenomena of nature taken as a totality. However, cosmology is not the science of all that exists, but the science which collects and orders natural phenomena into a whole, a unity called the "universe." The cosmological enterprise is thus closely related to astronomy and astrophysics, as well as to the physics of elementary particles.

The concept of totality is assumed in the idea of universe: all the components of the cosmos can be ordered as correlated parts of a *unique* system. Here lies a difficulty already acknowledged by Aristotle: science deals exclusively with the general, the universal; this exigency requires the possibility of repeating observations and experiences and of comparing their particular results. But the universe is unique, there is only one "observable" universe; moreover the human observer is "in" this universe from which he cannot escape.

As a result, the first and essential step in cosmology is the construction of a representation or model of the universe. Unavoidably, in this kind of representation or model of the universe what is supposed to be known is much larger than what is in fact known; in this sense, cosmology again approaches mythological narrative. In the last century, Auguste Comte wrote that, faced with the enormous complexity of natural phenomena, mankind begins by searching for reassurance in invented theories. Of course, science normally invents theories and builds models which exceed experience and which are presumed to allow predictions concerning the as yet unknown; in addition, science is totally dependent on the particular system of axioms assumed in each case. But no science requires such a large set of so daring assumptions as seem necessary to assemble the fundamental (mathematical) structure of the universe, a structure which must provide the framework for all other scientific constructions.

As Jacques Merleau-Ponty points out, cosmology was, at the end of the nineteenth century, still deemed an impossible science: "Cosmology was unmistakably excluded by Auguste Comte from the domain of the positive sciences, since, as was quite reasonable for his epoch, 'we do not even know if there is a universe,' meaning thereby that it was still unknown if all the cosmic systems observed or conjectured could be ordered in a unique system where all parts are in physical correlation."[1]   In fact, the beginning of modern cosmology can be traced back to 1917, when Einstein, trying to give general relativity all its extension and coherence, was led to state the problem in radically new terms. The search for a mathematical structure describing the universe, a cosmic geometry capable not only of encompassing the idea of unity, consubstantial with the idea of universe, but also of taking account of the axioms of general relativity, led Einstein to propose the first modern mathematical model of the universe.

This model assumes that knowledge of the physical universe is based on a system of fundamental laws formulated in mathematical equations relating the matter and energy content of the universe to the cosmic geometry. But mathematical laws as such cannot be verified, since the differential equations by which they are expressed describe elementary interactions taking place in an infinitesimal region of space-time. In order to establish the connection of these laws with descriptions of systems of finite dimensions and descriptions of the evolution of these systems in finite times, it is necessary not only to determine the numerical values of the parameters appearing in these equations, but also, and more importantly, to fix the initial and the border conditions, a central problem in cosmology.

As it is the case for every theoretical approach, the ultimate foundation of the "law" of nature is expressed in a set of *a priori* principles, that is, in a system of axioms. J. Merleau-Ponty writes: "The presentation of these specific principles, as provided by cosmologists, is a very delicate task, first of all because they are not always explicitly given, and maybe even not always perfectly conscious, and then because about these principles it is much more difficult to find a consensus."[2] The explicit listing of cosmology's specific principles is the task of the second part of this book.

In order to facilitate the reading of the following sections, which may appear a bit too technical to a reader completely unfamiliar with these subjects, we start with a brief outline of some of the key tenets of modern cosmology, theoretical as well as observational.

In 1905, physicists trying to explain the universe were facing significant problems. They had to admit the existence of an ether. This subtle incorporeal medium, whose existence was already assumed by the ancient Greeks, was supposed to permeate all space and to provide a material support to Newton's concept of absolute space. But this mysterious fluid remained beyond all experimental validation, even if the hypothesis of its existence allowed predictions liable to be tested by observation. Challenged by the failure of several experiments to prove the existence of the ether, scientists proposed new theories, grounded in turn on new axioms.

Thus, Fitzgerald and Lorentz put forward the hypothesis that all bodies in uniform and rectilinear movement must be affected by a contraction in their length in the direction of this movement. But experimental evidence did not support this idea: no shortening of moving bodies could be detected. In order to understand why this happened, a new postulate was advanced: the mass of a body in movement ought to change. Moreover, for each ("Galilean") frame of reference, a "local time" had to be defined. Lorentz and Poincaré calculated the relations between the coordinates of space and time for a system moving with uniform and rectilinear velocity relative to another one. However, the conception of an absolute space and time, then prevailing, concealed the deep meaning of these "Lorentz transformations." Thus, at that moment, everything was prepared, yet nothing was understood. The solution of that quandary constitutes the real achievement of Einstein's theory of relativity.

The problem Einstein faced stems from the fact that the concepts employed in these theories were tied to the most fundamental physical operation: the operation of measurement. But at that time scientists had not yet considered the possibility that, for moving bodies, definitions of spatial and temporal *measures* are not unambiguously stipulated, in so far as the basic measures of length of a rigid body or simultaneity between two separated events are dependent on the observer. Therefore, a radically new approach became inescapable.

a) Special Relativity

Einstein's special relativity is based on two fundamental postulates:

1) The principle of relativity, already found in Galileo's works,[3] states that no experiment can measure the absolute velocity of an observer. In addition, this postulate requires that the laws of physics should be mathematically equivalent for different (Galilean, i.e., unaccelerated) observers; the results of any experiment performed by an observer should not depend on his speed relative to other observers.

2) The universality of the speed of light, stating that the speed of light relative to any (unaccelerated) observer is the same regardless of

the motion of the light's source relative to the observer. It follows that special relativity singles out a class of preferred observers, called inertial or Galilean observers (or inertial or Galilean frames of reference): those which are unaccelerated.[4]

Thus Einstein could show that the spatial and the temporal coordinates of two inertial (Galilean) observers are physically related, and that the mathematical expression of this relation was precisely the Lorentz transformation, the meaning of which could now be understood. Furthermore, the admission of the fundamental postulate of the universality of the speed of light made it possible to put spatial and temporal measures on the same footing. By selecting new units, such that the speed of light called "c" is made equal to unity, time is measured in meters, and the ideal clock is a light ray traveling in a vacuum along a rigid rod.

Einstein's achievement made it possible to solve the problem confronting physicists at the beginning of the twentieth century:

i) According the postulate of relativity, all Galilean systems are equivalent for the purpose of the description of motion. Following the Galilean transformation,[5] in order to transform the observations of an observer in one such system to the observations of the same phenomenon in the other system, a vectorial addition of speeds results. As a consequence,

ii) the speed of light is differently measured in the two systems.

But, by postulate,

iii) the speed of light is a constant "c" in all Galilean systems, independent of the movement of the light's source.

As propositions ii) and iii) are obviously contradictory, it was necessary to replace these two postulates by the two Einsteinian ones mentioned above:

i) All Galilean inertial systems are equivalent; the formulation of the laws of physics is not dependent on the relative speed of the observers;

ii) Light propagates at constant speed "c" relatively to any Galilean system, irrespective of the motion of the light source.

In order to prove the consistency of these two postulates, Einstein showed how the notion of the simultaneity of occurrence of two separate events, as well as the notion of the length of a rigid body, were relative notions; as a matter of fact, he proved that the most fundamental measurement operations were not devoid of ambiguity! The consistency of these special relativity postulates (the equivalence of all Galilean observers and the universality of the speed of light) results from a radically new interpretation of the Lorentz transformation.

b) General relativity

Special relativity admits the equivalence of all Galilean observers in the formulation of the laws of physics, but singles out a preferential class of observers, the inertial (unaccelerated) observers, those not acted upon by an external force. In the case of accelerated observers (or frames of reference), or of observers under the effect of a gravitational field, a different approach has to be adopted.

On Earth at least, no inertial frame of reference can be found, since all "laboratories" are situated within the terrestrial gravitational field. In an inertial frame of reference (or laboratory), all particles at rest stay at rest; however, in a terrestrial laboratory a "free" particle falls. This led Einstein to formulate the principle of equivalence, on which general relativity is grounded: the identity between inertial and gravitational mass.

In the fundamental (Newtonian) law of dynamics

$$F = m \cdot a$$

the inertial mass m appears, the mass which, so to speak, resists acceleration.

The gravitational mass M appears in Newton's law of gravitation:

$$F = G \cdot M M'/r^2$$

The gravitational action (attraction) of the mass M' (for instance, the Earth) on the test particle, whose mass is M (for instance, Newton's famous apple), will accelerate the test particle to

$$a = F / m = - G \cdot (M/m) \cdot M' / r^2$$

As Einstein noticed, experiment had shown that in the vacuum all test particles fall with the same acceleration.[6] Thus, in a gravitational field, the acceleration GM'/r^2 is independent of the test particle's composition and mass: the value M/m is the same for all bodies. If appropriate unities are chosen, this value can be set to unity; consequently, the gravitational mass M is identical to the inertial mass m.

This identity of gravitational and inertial masses makes it impossible to distinguish locally between inertial forces of accelerated systems and gravitational forces due to a gravitational field. The *principle of equivalence between gravity and acceleration* can accordingly be formulated in these terms:

Locally, in a small region of space, uniform gravitational fields are equivalent to frames of reference that accelerate uniformly relative to inertial frames of reference.[7]

Stated in this way, an accelerated system can be considered similar to some particular inertial system, a system freely falling, when placed in a suitably chosen gravitational field; the classical example is Einstein's "freely falling elevator." But the equivalence principle applies only *locally*. Consequently, there is no *global* reference frame which is everywhere freely falling; to construct a global inertial frame within a gravitational field (such as the gravitational field of the Earth) is impossible. In order to make the laws of physics independent of the laboratory, to make them coordinate invariant, Einstein was led to formulate the *principle of general relativity*:

All frames of reference, whatever their motion, should be equivalent for the mathematical formulation of the laws of physics.

The concept of "free observer" is modified by this principle. The principle stating that all reference systems are equivalent is based on the conditions substantiating the Lorentz transformation. However, because the above mentioned equivalence principle applies only locally, to encompass all space by one unique ("absolute" in Newton's sense) inertial (Lorentz) frame of reference is impossible in a gravitational field. Consequently, the postulate had to be narrowed: the laws of physics must be the same in any *local* Lorentz frame of reference.[8] One thus places a Lorentz frame on some small part of space, another in the immediately adjacent part, and so on until all space is covered. But since it is impossible to construct, in a gravitational field, a *global* inertial reference frame, the entire structure of space and time must be revised. Instead of the Euclidean structure of space and time, a non-Euclidean geometric object, called "space-time" must be introduced, whose properties only observation can elucidate.

Einstein succeeded in setting up a system of equations capable of correlating the matter and energy content of the universe with the geometrical structure of space-time. The solution of this system of 10 non-linear differential equations makes it possible to find the metric – the "curvature" – of space-time, a curvature precisely generated by the matter/energy distribution of the universe. Their solution requires the previous determination of the relevant initial values and the border conditions.

The search for a satisfactory cosmological model is hindered by three main obstacles:

i) The mathematical complexity of Einstein's system of equations: no unique solution giving "the" curvature of the universe exists;

several are possible. Thus many different universes are conceivable.

ii) The difficulty in attaining a sufficiently accurate observational determination of the quantity, the composition, the properties, and the distribution of the matter and energy content of the universe.

iii) The difficulty stemming from the fact that the theory predicts that only a small part of the universe is accessible to the human observer, a part that may be vanishingly small.

In spite of these problems, the development of a sophisticated "Big Bang" cosmology was not precluded, as the ensuing exposition, which does not follow its historical unfolding, shows. Since we concentrate our attention on the "problem of scientific knowledge," we will, as in our examination of Plato's *Timaeus*, present the modern cosmological model from a viewpoint relevant to our purpose. This means that we will enumerate the fundamental underlying presuppositions (axioms) of the model. Once this axiomatic system has been expounded in detail, we will proceed to establish the connection of the model so constructed with the attainable observational facts.[9]

A) The Standard Big Bang Model: A Short Description

In this section, a mathematical model will be described, a model which aims to represent the universe, and to reveal what orders and unifies natural phenomena into one totality. Such a model, as will soon become apparent, requires a particularly sophisticated mathematical instrumentation; it is based on Einstein's general relativity, and represents the state of the question around the years 1940-1950. Elaborate additions, such as the inflationary scenario which will be briefly alluded to further on, may be considered as recent developments of the model (1981); today, the frontier of cosmological research centers on Grand Unified Theories, quantum gravity, super strings, and "theories of everything."

But first this question: what is a mathematical model?

Broadly speaking, a mathematical model is the conjunction of 1) a specific, well defined set of assumptions (axiomatic propositions);[10] generally, each model possesses its particular set, and 2) the total collection of mathematical knowledge available. Since Mathematics – with a capital M – as well as Logic – with a capital L – can in principle both be applied to every scientific model, what distinguishes one particular model from another is the set of their specific axiomatic propositions. This was already clear to the Greek philosophers. Aristotle refers to this subject in the *Posterior Analytics*:

> A science is one if it is concerned with a single genus or class of objects which are composed of the primary elements of that genus [i.e., the principles]. . . . One science is different from another if their principles do not belong to the same genus, or if the principles of the one are not derived from the principles of the other. (*Posterior Analytics* I 28, 87a38-b1)

In our definition of a mathematical model, another Aristotelian idea is implicit. If the first principles of a model are given as premises, the entire set of theorems mathematically or logically deducible from these premises, that is to say everything the model is able to give a description of, is already known. In other words, the totality of the deducible theorems is contained, potentially, in the system of axiomatic propositions founding the model. (cf. Aristotle, *Posterior Analytics* I 24, 86 a 22-25).

Thus two mathematical models are indistinguishable if the fundamental hypotheses of one can be deduced from those of the other, if the system of axiomatic propositions of one is already contained in the other.

Consistent with the method sketched in the first part, we will now review, in as much detail as required by the intended analysis, the presuppositions on which the cosmological model named "standard Big Bang" is founded. This task will be carried out stepwise: the different components of the model will be added one after the other, until the essential set of axioms on which the standard Big Bang model is grounded is completed.

The "philosophical" presuppositions

Any construction of a mathematical model rests inevitably on a certain number of philosophical presuppositions – one might also call them "prejudices." Such assumptions are not always explicitly formulated, and most of them have been passionately debated for centuries. Our claim is not to produce the complete list of these philosophical presuppositions; only the subset of the most interesting assumptions shared by both models analyzed in this book will be given.

• An observer-independent reality does exist; the complete set of data that can possibly be collected from this observer-independent reality is called the "universe." The observer himself is however part of this universe.
• The subject/object relation between a human observer and such an external reality does not entail a loss of crucial information. In other words, the segmentation of the universe into a subject and its object is a natural one, and, when assumed, entails no irreversible destruction of information about the external reality. A universe considered to be some kind of *continuum*, such as postulated by Parmenides, is excluded by this assumption.
• For a human observer,[11] sense perception provides the only access to reality; this means that only *measurement* allows a scientific approach of this external reality. Actually, the scope of this assumption is larger, since it includes, as taken for granted, that the appearances of the objects perceived are sufficiently *stable* to be distinguished and named (cf. *Timaeus* 49d-50a). To discern correlations between these objects (or between different states of one object), such a stability is *a priori* necessary.
• Finally, pure reason, the intellectual activity which in human beings operates in total independence from any sense-perception, is apt

to reach a certain level of knowledge or truth. Human beings thus have the capacity to formulate universal principles, based on such abstract notions as harmony, beauty, symmetry, causality, necessity, etc.

The geometric axioms

In the following sections, the presuppositions *specific* to the standard Big Bang model will be listed. The construction of the model starts from the simplest assumption. Then, step by step, the required structure is built, until Einstein's field equations are reached. As it is extremely difficult to find the adequate solution for these equations, the process has to be reiterated, with the effect that drastic simplifications of the universe are now assumed to hold. These simplifications finally lead to the Friedmann-Robertson-Walker metric function. Only at this stage is the model ready to be tested by observation, only at this stage can the model be subjected to verification or falsification.

First, a succinct list of the primordial geometric axioms on which the model is based will be given. They originate in Einstein's fundamental idea, according to which the universe is a geometric object: space-time.[12] In this way the universe *can* be geometrically defined.

The main feature of a geometric object is *neighborhood*, a rather complex notion, close to the idea of infinite proximity. Neighborhood is an essential property of objects,[13] that particular property which allows them to be qualified as "geometric" when, after subtracting all other essential or contingent qualities, only this one is retained. The concept of neighborhood correlations between geometrical objects allows us to define continuous transformations as those which leave such correlations intact.[14] Modern set theory has considerably extended and deepened this idea of a geometric object.

In the geometric approach to Big Bang cosmology, the universe is identified with an arbitrary set whose elements are points of a *n-dimensional topological space*, the reason being that such a topological space may be endowed with the necessary attributes permitting

continuous mathematical transformations to be
axiom of the model is:

Axiom B1

The universe is a geometric object, and that geometric object is a topological space.

The points of this space are called "events." Each event is unambiguously determined by its space-time coordinates; each event happens "here and now," (*hic et nunc*). The universe is thus defined as the totality of physical events. The important consequence of this assumption is to establish the independence of the universe from any "exterior" frame of reference. In this topological space, physical events are identified by their *name*: "the event happening here and now."

Events, i.e., the points of the topological space modeling the universe, ought to remain distinct and separate; otherwise, the concept of a well-defined physical event is meaningless. To achieve this end, a new axiom, named a "separation axiom" has to be postulated. There exist several separation axioms; the one chosen is:

Axiom B2

The standard model postulates that the event/points of the topological space representing the universe satisfy the Hausdorff separation axiom.

A Hausdorff topological space is a space such that for each pair x, y of distinct points of that space, there exist disjoint neighborhoods of x and y.

Axioms B1 & B2

The universe is modeled by a Hausdorff topological space.

It is admitted that the universe does not present isolated or disconnected "portions" between which information conveying causal information cannot flow. Actually, should the universe have such disconnected portions, we would be forever unable to know them; thus it seems adequate to rule them out axiomatically. This requires that the topological space representing the universe be connected.[15] Furthermore, one also rejects out of hand any possibility of a universe having "edges" or some kind of boundary; anyway, what could *be* beyond such boundary? This last condition is not to be confounded with an infinite size; for instance, a spherical surface is finite, but has no edges nor boundary. We thus add

Axiom B3

The universe is connected and without boundary.

The next step follows naturally. Mathematics being the primary tool of physics, the topological space modeling the universe must be provided with precisely those properties which allow us to obtain a knowledge of the universe by means of the mathematics currently best understood. Differential equations are the most suitable mathematical tool for the description and analysis of dynamic phenomena, that is, the description of the temporal change of physical situations. To assure the possibility of giving a meaningful definition of differential functions in the universe, additional conditions must be stipulated: the universe must be "smooth" everywhere. This condition, expressed mathematically, is

Axiom B4

The geometric model representing the universe is a "differentiable manifold."

A differentiable manifold is essentially a smooth topological space, such that an infinitesimally small region around each point of this space resembles an Euclidean space. In this space, the parameters characterizing it change smoothly and *continuously*.[16] Moreover, the number of independent parameters of the manifold are called the "dimension of the manifold." Taking the surface of a sphere as an example, we see that two independent parameters suffice to identify each point, viz. the two polar coordinates which may be named latitude and longitude, or θ and ϕ. This surface is smooth, and at each point a "tangent" two-dimensional Euclidean space – the tangent plane – can be conceived. Thus the surface of the sphere is a two-dimensional differentiable manifold.

Mathematically, the operation associating a set of points (of a manifold) with the parameters characterizing them, is equivalent to the "mapping"[17] of the points of the manifold onto an Euclidean space, the dimensions of the Euclidean space being the same as the number of independent parameters of the manifold. For instance, the Earth, being approximately spherical, and which is everywhere curved, can nevertheless be mapped continuously into a set of two-dimensional Euclidean spaces: our usual maps, atlases or charts. A differentiable manifold is thus everywhere locally Euclidean; such a manifold is smooth, continuous and can be described by a number of independent parameters equal to its dimensions. On a larger scale, the topology of a n-dimensional manifold may look entirely different from an Euclidean space (as shown, for instance, by the round Earth and the flat maps we utilize); but locally, in every infinitesimally small part of its "surface" the manifold can be mapped onto the tangent "plane" at that point. Thus, at each point of the manifold, local charts (atlases) can be established but no unique chart can "cover" the entire manifold. In fact, globally *any* topology is possible, as long as the manifold stays smooth.

Having thus defined the differentiable manifold representing the universe, the next step is to assign to it its dimension. Based on every day experience, it is postulated that:

Axiom B5

Four dimensions suffice to specify any point/event of the manifold: three spatial parameters and one temporal.

This manifold is therefore called "space-time."

Axioms B1 through B5

The universe is modeled by a 4-dimensional differentiable manifold, satisfying the Hausdorff separation axiom, connected and without boundaries.

The universe is thus modeled by a 4-dimensional geometric object. Besides the fact that such a model is particularly difficult to visualize – it is very hard to imagine a world of *four* dimensions – what is most surprising in this approach is that time is considered to be one of its coordinates. In a way, everything is "frozen" in this universe: nothing ever changes, everything *is* and nothing ever becomes. An example often used to illustrate this situation is a reel of film. The film never changes, but the illusion of change can be accomplished in the usual way. Note however that in the film, "time" is also measured in meters (or number of frames).

These conditions are still insufficient to assure that the mathematics we know how to use can be applied as required. We have to add the condition of the convergence of the sums used to define multiple integrals. This is equivalent to the requirement that each point of the manifold be covered by only a finite number of charts. Mathematically this is expressed in these terms.

Axiom B6

The manifold is paracompact.

Adding still more conditions to assure the suitable differentiability of the manifold, we arrive at:

Axioms B1 through B6

The universe is modeled by a four-dimensional Hausdorff C^∞ manifold, connected, without boundaries and para-compact.

The next step consists in providing the topological space – space-time – with a metric function. A topological space possessing everywhere a metric structure is called a "metric space." The definition of the metric function *d* is as follows:

Let *E* be a set. To each couple *x* and *y* of *E* is assigned a real number $d(x,y)$, called the "distance" between *x* and *y*, satisfying the following conditions:

i) $d(x,y) \geq 0$;
ii) $d(x,y) = 0$, *if and only if* $x = y$;
iii) $d(x,y) = d(y,x)$, *for all x and y in E* ;
iv) $d(x,z) \leq d(x,y) + d(y,z)$, *for all x, y, z of E.*

Axiom B7

The topological space by which the universe is modeled is a metric space.

The intuitive idea of distance can be applied, not only to Euclidean spaces, but to any n-dimensional manifold as well. It should be understood that adding a metric function to a manifold is equivalent to providing this manifold with a structure; otherwise the manifold is but a shapeless, amorphous, collection of points. The metric function defines the curvature of the manifold and, without a metric function, no unambiguous measurement can be performed on the manifold.

The paracompactness of the manifold established in axiom B6 is equivalent to the assumption that the manifold admits some positive-definite metric, meaning that the distance function $d(x,y) \geq 0$ for all x, y. That is, the distance is always either positive or null, and two points at null distance are one and the same point. In general relativity, however, the metric function is defined as "pseudo-Riemannian." Such a function admits that the distance function $d(x,y)$ can take positive, null or negative values. It thus becomes possible to define curves, the distinct points of which are separated by an invariant distance equal to zero. Although the distance between two such points is zero, they are not the same point, they remain distinct points. In general relativity, such null curves are precisely the paths followed by light rays in the four-dimensional space-time manifold thus defined. The path of a test particle in general relativity's space-time is called a "geodesic;" if this test particle is light, then this particle is called a "photon." Photons travel at constant speed equal c (or equal 1 in units of general relativity) in any frame of reference, and the equation specifying null-geodesics are the equations of the displacement of light-photons. Photons, and light, occupy a central place in general relativity: it is postulated that nothing can exceed the speed of light. Specifically, no causal information can be transmitted in the universe faster than light. This value being finite, a limit is established beyond which all causal interaction is forbidden. This is postulated in:

Axiom B8

No information (or causal action) can be transmitted faster than the speed of light, the value of which is constant and finite.

Although time is just one of the four parameters defining space-time, the fact that time flows incessantly and determines causality must be taken into account. This special character of time is the reason why a pseudo-Riemannian metric, instead of a positive-definite metric, is chosen. A pseudo-Riemannian metric is such that the manifold acquires everywhere a "Lorentz signature" a condition more restrictive than axiom B6, the paracompactness of space-time; any manifold with a Lorentz signature is also paracompact. A Lorentz signature means that, everywhere in the manifold, at every local event/point, a Lorentz frame of reference does exist.[18] This postulate implies that one can always find, locally, the results of Einstein's special relativity, and therefore also those of Newton's physics.[19]

The Lorentz signature is *(-1, +1, +1, +1)*, meaning that the distance function in flat space-time is of the form:

$$(ds)^2 = - (cdt)^2 + (dx)^2 + (dy)^2 + (dz)^2$$

In Euclidean geometry, when the distance between two points is zero, the two points must be the same point. In the geometry of general relativity, in a manifold presenting everywhere a Lorentz signature, due to the minus sign of *(cdt)²*, when their interval *(ds)²* is zero, two events may be on two different galaxies, but they are nevertheless separated by a null ray (a light ray). Technically, this is expressed by assuming that the metric function of the space-time manifold has the general four- dimensional form :

[equation 1]

$$(ds)^2 = g_{ij}\, dx^i\, dx^j, \quad (i,j = 0, 1, 2, 3), \quad \text{with}$$

diag *(-1, +1, +1, +1)*, in local Lorentz coordinates. Since *locally* space-time is, by axiom, flat, this signature, the guarantee of a simple geometry, is everywhere *locally* true.

The metric function g_{ij} determines the (four-dimensional space-time) "interval" between any two events and the curvature of space-time.

From all this it ensues that we have to add the following axiom to the construction of the model:

Axiom B9

Everywhere in the manifold modeling the universe there is defined a Lorentz signature metric.

We now summarize the standard model's axioms listed up to this point:

Axioms B1 through B9

The universe is modeled by a Hausdorff separable, C^∞ four-dimensional manifold, connected and without boundary. Everywhere on this manifold, a Lorentz signature metric function can be defined.

Einstein's gravitation

In the preceding section, the *geometrical* presuppositions (axiomatic propositions) on which the Big Bang model is grounded were briefly enumerated. This model shows the universe to be a precise geometrical object, built to allow the analysis of physical phenomena by means of currently known mathematics. In other words, if the description of the universe presented by this model is realistic, then contemporary mathematics is the adequate instrument for acquiring scientific knowledge about the universe. However, the model is still incomplete. In the next stage, the postulates specific to Einstein's theory of gravitation must be introduced by presenting its indispensable axiomatic propositions.

Introductory remarks

The procedure followed in the building of the Big Bang model is in many aspects parallel to the method employed by Plato in the *Timaeus*, as described in the first part of this book.

Both approaches postulate a universe exhibiting properties of simplicity and symmetry which allow the use of contemporary mathematics, and both try, through their application, to achieve a verisimilar knowledge of physical phenomena. By postulating axioms T10, T11 and T12, Plato in the *Timaeus* assumes that a benevolent demiurge has, as far as possible, imposed upon the universe beauty and harmony, abstract properties which, it is claimed, can be mathematically formulated. Moreover, Plato draws the consequences of the axioms he admitted: since the celestial bodies are the most perfect beings, their motion *must* be the simplest and the most harmonious (i.e., the most *symmetric*), that is to say circular. The model of the universe built according to this axiom must include this solution; consequently, the world-soul in the *Timaeus* is manufactured as an armillary sphere. In Plato's time, mathematics were basically restricted to the geometry to which Euclid, some years later, was to give its final formulation, and to the arithmetic of proportions. On the other hand, it had already been discovered at the time that certain fixed arithmetical proportions could adequately describe acoustic phenomena, and this insight led Plato boldly to extrapolate from musical harmony to a mathematical explanatory principle applicable to the whole universe.

This line of thought, which can truly said to be "revolutionary" is mirrored in the Big Bang model. By axiom T3, Plato did not only postulate causality in the universe, but he went further: those causes of all that becomes in the universe under the "governing" mathematical rule of the world-soul, are also "reasonable," "logical," and therefore accessible to human capabilities. Causal explanations in the universe – if and where such explications exist – should be based on mathematics human scientists are supposed to be able to develop, even if such explanations are at best "verisimilar" (axioms T11 and T12).

Plato's insight lies at the roots of probably the most fundamental of all *a priori* assumptions admitted by modern science: *nihil est sine ratione*, so formulated by Leibniz, who called it *principium grande, magnum et nobilissimum*. According to this "un-Platonic" principle, *everything* in the universe, including the universe itself, must possess a sufficient reason for its existence, and its being just as it is. In Leibniz' words: "We consider that no fact will be true, nor exist, nor will any proposition be truthful, without a sufficient reason why it is so and not otherwise. Even if in most cases these reasons will remain unknown to us."[20]

This means not only that everything which becomes must of necessity become owing to some cause (*Timaeus* 28a; cf. axiom T3, a law which only establishes a rule of precedence), but that, in addition, within the best of all possible worlds, ruled by a "pre-established harmony," this cause is in conformity with reason. Here again Plato's intuition was astonishing: the Platonic universe is far from being the best of all possible universes, since in the *Timaeus* it is stated that the "reasonable" causes, those that can be mathematically formulated, are hampered by the "errant cause," the *anagkê* of axiom T8. As the demiurge is not all-powerful, reason does not pervade the universe but only goes "as far as possible," and an undetermined portion of the universe entirely escapes the mathematical domination of the world-soul.

Its vast scope notwithstanding, this *principium grande, magnum et nobilissimum* holds a central position in modern science. One can hardly imagine scientists, Einstein for instance, building mathematical models encompassing the entire cosmos, unless they accept at least some version of this principle.[21]

In summary, modern cosmology basically follows the guidelines set up in the *Timaeus*. Once they have accepted that mathematics, the mathematics of differential equations, give an adequate description of a great number of physical phenomena, the proponents of the Big Bang model extrapolate, as audaciously as did Plato in the *Timaeus*, and provide the universe with the appropriate fundamental mathematical structure, but in this case the *entire* universe becomes accessible to human understanding and investigation.

Einstein's dynamic axioms: Matter and energy in the universe

So far, we have described the topological structure assumed to model the geometry of the universe. On this basis, Einstein now established the relation between this geometry and the matter and energy content of the universe, following this daring idea: curvature in the geometry of space-time manifests itself as gravitation.

As the equivalence between matter and energy is established by Einstein's famous equation

$$E = m \bullet c^2 \quad \textbf{[equation 2]},$$

new axioms are required if a universal validity is to be assigned to the relationship between geometry and matter/energy. Conceivably the most revolutionary idea Einstein formulated with regard to general relativity is that there exists a dynamic connection between such disparate entities as are the geometry and the matter content of the universe, a connection that can be precisely formulated mathematically. "The effect of geometry on matter is what we mean today by that old [i.e., Newtonian] word 'gravitation,' and matter in turn warps geometry. ... Space acts on matter, telling it how to move. In turn, matter reacts back on space, telling it how to curve."[22] It is precisely this close connection between matter and space-time that is at the root of the difficulties presented by this mathematical picture of the universe. The matter/energy content of the universe, a moving body for instance, affects space-time which curves it, but curved space-time affects the way bodies move in the first place, and so on ..., a hopeless enterprise, unless a mathematical tool is found capable of picturing this permanent mutual inter-relationship.[23]

Such a tool are Einstein's field equations. These equations are a *tensor* system of partial differential equations relating the metric of the manifold, also represented by a tensor, to the tensor representing the energy distribution of space-time. This energy distribution is taken as the source of the gravitational field and is phenomenologically described by the energy/momentum tensor.

First, we summarize Einstein's key assumptions, adding these two new axioms:

Axiom B10

The dynamical equations describing gravitation are such that they involve the matter/energy densities of the universe only through the energy/momentum tensor.

Axiom B11

The general form of the dynamical equations is a set of tensor equations of the form

$$E = 8 \pi G \cdot T \quad \text{[equation 3]},$$

where G is the gravitational constant, T the energy/momentum tensor and E another tensor depending only on the geometrical structure of the space-time manifold and its metric function.

These tensor equations [3] are the Einstein field equations, the basic piece in the construction of general relativity. E is the Einstein tensor that is always generated directly by the local distribution of matter, whereas T is a geometrical object called the energy/momentum tensor which acts as the source of gravity in the universe. No coordinates are needed to define E and none are needed to define T. Locality and coordinate-independence are the key features of the field equations.

These axioms embody Einstein's fundamental idea that the laws of physics should not be dependent on the frame of reference chosen for their description. Mathematically, this can only be achieved through a tensorial approach. The dynamic laws prevailing in the universe must therefore take a tensorial, coordinate system-invariant form. In this way, the principle of scientific objectivity, known as the *principle of invariance*, finds its mathematical expression in the model. In effect, the tensorial formalism allows one to consider space-time locally, according to Einstein's dictum that physics look simple only when analyzed locally. Everything in Einstein's theory happens locally, and this explains why no coordinate system is required in the definition of the tensors *E* and *T*. This idea is formalized by axiom B9, according to which, at every event/point of the manifold, a local inertial (i.e., Lorentz) frame of reference exists. On the other hand, there is an overall structure of the theory, and this attribute can manifest itself locally. The local manifestations of this overall structure are the curvature of space-time and the resulting gravitational forces.[24]

Furthermore, this tensorial structure allows automatically to reduce Einstein's formalism to the well tested Newtonian theory in the case of weak gravitational fields, as is evidently desirable.

In fact, the Einstein field equations given by

$$E = 8 \pi G \cdot T \quad \textbf{[equation 3]}$$

"show how the stress/energy of matter generates an average curvature *E* in its neighborhood. Simultaneously, the field equations are a propagation equation for the remaining, anisotropic part of the curvature: they govern the external space-time curvature of a static source (Earth); they govern the generation of gravitational waves (ripples in curvature of space-time) by stress/energy in motion; and they govern the propagation of those waves through the universe. The field equations even contain within themselves the equation of motion (Force = mass • acceleration) for the matter whose stress/energy generates the curvature. The field equations govern the motion of the planets in the solar system; the deflection of light by the sun; the collapse of a star to form a black hole; they determine uniquely the external space-time geometry of a black hole; the evolution of space-time at the end point of collapse; the expansion and recontraction of the universe; and more, and more."[25]

Time and causality in the universe

In spite of the great power of these field equations, the construction of the axiomatic structure of the cosmological model is far from complete. New axioms are needed to take into account other features of the universe. First, the space-time manifold must be endowed with properties bringing the dimension time into agreement with the intuitive notion of time. Thus, a one-way directionality is allotted to this dimension, and the so called "arrow of time," pointing irreversibly from past-to-present-to-future, is introduced.

In the relativistic approach, this "common sense" assumption has nevertheless surprising consequences. At *every* point of the manifold, a local inertial (Lorentz) frame of reference is assumed to exist in which geometry is simple. This presupposition means: in any particular local frame of reference, observations can be made recording the location (x, y, z) and the time (t) of some event in such a way that the interval between (x_1, y_1, z_1) and (x_2, y_2, z_2) is independent of time; the clocks at every point are synchronized and run at the same rate; and the geometry of space is Euclidean. If the postulate of the universality of the speed of light is added,[26] the interval

[equation 4]

$$(ds)^2 = - (cdt)^2 + (dx)^2 + (dy)^2 + (dz)^2$$

is the same for any inertial frame. (Here c is the speed of light; notice also the Lorentz signature $-1, +1 +1, +1$.)

The interval $(ds)^2$ in equation [4] may turn out to be positive, negative, or equal to zero. If the interval is positive, then the events so separated are said to be spacelike separated and no causal interaction can exist between two so separated events.[27] If the interval is negative, one event is said to be located within the light cone of the other and, depending on the arbitrarily fixed direction of the arrow of time, one will be situated in the future light cone of the other (or, conversely, in its past light cone). Finally, the zero interval defines events with lightlike or null separation. All this is illustrated in Figure 2.14, drawn taking the value $c = 1$, and where for clarity's sake, only two spatial dimensions, x and y, are depicted.

Figure 2.14

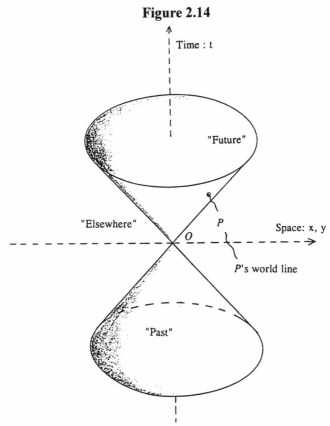

According to this figure, the universe is split into three non-overlapping zones from the point of view of some particular event O, of some particular *hic et nunc*: its past light cone represents all the events from which causal effects and information may be received; while its future light cone defines all physical events on which this particular *hic et nunc* may have a causal effect, to which it may send information; and, outside these two cones, there is an "elsewhere" a spacelike separated region of the universe, causally disconnected from this particular point, from which no information can be received, and to which no information can be sent. The track through space-time of one event is called its "worldline" or its "geodesic." At each point of a geodesic the event is described as being somewhere in space and somewhen in time. The tracks of light are the null geodesics, they follow exactly the cone itself. Distinct points on a null-geodesic are separated by a zero interval, i.e., their interval $(ds)^2$ in equation [4] equals zero. Furthermore, events such as P lying outside the light cone of the "here and now O", can eventually enter the future light cone of O and causally interact with it "then," if their respective geodesics should meet.[28]

Two additional comments on Figure 2.14 are apposite. First, we can only observe what lies in our particular past light cone from our particular *hic et nunc* on the solar system; and that may turn out to be only a minimal fraction of the universe. Secondly, we have unambiguously defined the (arbitrary) past-present-future arrow of time only in a local inertial frame of reference. However, such a classification cannot be done in a continuous way on the whole manifold. Depending on the matter/energy distribution, and its resulting topology according to the field equations, it may be possible to have at one point a certain past-future orientation and, starting out from that point/event in a closed path, to return to it, having the time direction reversed. Such would be the case if some parts of the topology of the universe resemble a Moebius band. Evidently, in the construction of the cosmological model such a contradiction to our prejudices is considered to be unacceptable; it is eliminated by adding one more geometrical axiom to the list:

Axiom B12

The topological manifold representing the universe must be time-orientable.

This means that at every event/point of space-time, in each particular Lorentz frame of reference, a distinct choice has been made as to which light cone is the future and which is the past cone, and moreover that this choice is continuous from event to event throughout space-time.

We are still far from a complete list of axioms. The next group of axioms will introduce the idea of causality in the universe, in a manner reminiscent of *Timaeus* 28a.[29]

In fact, the above axiomatically introduced time-orientability is not itself a strong enough condition to assure causality in the usual sense. In general relativity, causality is taken as the result of a signal emission from one event (the "cause") to another (the "effect") in a time-oriented universe. The fastest signal is transmitted along time-like (or null) pathways (geodesics).[30] Now, the form which these causality-carrying pathways take in the four-dimensional manifold representing the universe is dependent on the intricate relationship between the matter/energy content of the universe and the geometrical properties of the topological structure resulting from Einstein's field equations (things appear simple only locally!). There is nothing, however, to prevent these pathways through space-time from forming loops, so as to pass more than once through the same point (or infinitely nearby). The result, exploited in science-fiction, allows one event to act causally on its own past.

To save the essential notion of causality, one further postulate concerning the mathematical structure of the manifold representing the universe should be added. Formulated in 1969 by S. Hawking, it is called the "causal stability condition": a manifold M with metric g is said to be stably causal if there exists a (nowhere vanishing) time-like vector field t such that the Lorentz metric on M admits no closed time-like curves.

This definition introduces a new axiom.

Axiom B13

In the universe, the condition of causal stability obtains.

The causal stability condition requires that in the universe, at every point in space-time, a strictly increasing "time" flowing only from past to future can be defined. But this time is not conventional time, because the postulated time function is not unique and because the space-like three-dimensional "hyper-surface," defined by the condition $t = constant$ are not necessarily connected.

Here a new concept is introduced: the possibility of constructing space-like three-dimensional "slices" of the four-dimensional universe-manifold, i.e., three-dimensional sub-manifolds representing "all space at an instant of time." Do all types of manifolds so far considered admit this form of slicing? Fortunately, those which are stably causal do. And, furthermore, such manifolds possess "achronal" slices, that is, slices where no point precedes (in the sense of the flow of time) any other point. Precedence among events refers to the capacity of some event to send messages to another one; two events can thus be called respectively "cause" and "effect."

Solving Einstein's field equations

The next problem is essentially mathematical. Einstein's field equations are a set of partial differential equations, which describe the dynamic interrelation between geometry and matter, but not observable phenomena, which are described by the *solutions* of these equations. To be able to find these solutions, the equations must be integrated and this, in turn, requires the complete set of initial and boundary conditions. How can this be accomplished? It can be attained if and only if the topological structure featured by the universe is such that the information which may be gathered on one particular achronal hypersurface is sufficient to determine the entire manifold. This is important because human observers have access to only *one* achronal hypersurface, the one determined by our pointlike *hic et nunc*;

cosmologically speaking, all of human history is but a point on the time axis, and the entire Milky Way is but a point in space.

Mathematically, and omitting many details, this additional postulate is expressed as a new axiom:

Axiom B14

On the space-time manifold representing the universe exists an achronal slice such as S, said to be a "Cauchy surface" of the manifold, having the property that every point in the manifold is in the domain of dependence of S.

The domains of dependence referred to in this axiom are precisely defined so as to separate event precedences. It follows that in the space-time manifold there exists an achronal surface (Cauchy surface) such that the information on this particular slice is sufficient for the determination of the physics of the entire manifold.

To postulate a Cauchy surface (axiom B14) is a strong condition to impose on a manifold. Any manifold with a Cauchy surface is stably causal. The time function referred to above can now be chosen in such a way that any slice at $t = constant$ will also be a Cauchy surface. But it must be underlined that the existence of a Cauchy surface in the universe-manifold is not implied by the theory; it must be postulated as an additional requirement.

We close this section on Einstein's gravitation by a brief reference to the complex problem of singularities in space-time, which seem to emerge unavoidably in the solutions of Einstein's field equations. A singularity, in this context, is a point in space-time where certain parameters take unbounded, that is to say, infinite, values and where the geometrical approach breaks down. But, by definition, a manifold is necessarily smooth. This means that singularities must lie outside the manifold. What happens when, in a certain space-time, a time-like geodesic, for instance the trajectory through the manifold of a freely falling body, comes suddenly to an end? The body cannot disappear. And if it does, then where does it go? Einstein's system of field equations accept a number of different solutions, in which such kind of singularities appear. In fact, a space-time in Einstein's theory will in most cases show, under a rather general set of assumptions, singularities of some kind.

One such singularity occurs in the neighborhood of very strong gravitational sources. There the curved geometry may have the property of "dragging" all outgoing light rays back into the source. Since nothing can travel faster than light, all matter and energy within a 3-dimensional hyper-surface (the "horizon") is trapped and so gives rise to a singularity. Stars of a certain type, after exhausting their nuclear fuel, collapse into an extremely reduced volume: a black hole singularity may arise in this way. But, of course, the most famous singularity is the Big Bang itself.

Singularities, then, can occur in space-time as a consequence of the geometrical axioms of the model.[31] They are not directly observable, but they do exert an influence on their surroundings that may, in principle, be detectable.[32] However, no known physical theory can describe what happens at the singularity itself.

Here ends our exposition of the geometrical and dynamic postulates required in order to apply Einstein's theory of gravitation to the universe, and which constitute the basic axiom system of the standard Big Bang model.

This construction was guided by two fundamental objectives:

1. To allow the mathematics we know how to use to be applicable; and

2. To give to the concepts of time, of space, and of causality a geometrical structure compatible with human prejudices. And this, basically, by strictly following the logical postulates originally formulated in Aristotle's *Organon*.

However, in order to make sense out of what the astronomical observations reveal about the universe, one has still to establish the relation of all the enumerated axioms with this reality, a task which requires the prior analysis of several other questions.

Isotropy and homogeneity of the energy/matter content of the universe: the standard Friedmann-Robertson-Walker model

In Einstein's general relativity the geometry of the universe cannot be discerned unless the energy/momentum tensor T, appearing on the right hand side of the field equations, has been elucidated. So just as the Einstein tensor E is deemed to incorporate the geometrical features of the universe, in a similar manner the energy/momentum tensor should mathematically represent its matter/energy content. Its construction ought to follow from experimentally gathered, empirical, evidence. And, ideally, one may wish to know from separate, independent, sets of observations the physical values which determine both tensors.

What can actually be observationally inferred in the universe is the subject-matter of the following section. Here we first enunciate the next axiom:

Axiom B15

The mathematical laws that have been found to apply to physical phenomena, as elucidated up to the present date by solar system-bound scientists, are valid everywhere and everywhen throughout the universe.

Observational evidence has so far given no reason to doubt this claim and, moreover, cosmology would be an impossible task without this assumption.

A second, radical, assumption is equally essential for a cosmological science to exist:

Axiom B16

Even if its size is gigantic, the universe is supposed to have a simple structure.

The claim that these two axiomatic propositions are not contradicted by the available observational evidence is a matter of some debate. The fact that the universe is enormous relative to human scales is, of course,[33] commonly accepted, although some models of "small" universes have been proposed.[34] But the assumption that the universe is simple is not evident, unless Leibniz' principle is taken for granted.[35] On the other hand, if one takes into account that the universe is very large, that the speed of light is finite, that the pace at which astronomical events generally proceed is equally finite, and that in general the life span of most astrophysical objects and processes is very long, compared to the 5,000 years of human history, it may be legitimate to assume that cosmologically relevant things change so slowly that the universe appears *to us*, at least dynamically, to be "simple."

To build a model of the universe which is "simple," and in which the known laws of physics obtain everywhere (conditions requested by axioms B15 and B16), additional drastic simplifications must be introduced. As a result, the Einstein field equations will become mathematically more tractable,[36] a necessary requisite to obtain a model that can be correlated with observation. These axioms extrapolate the vastness of the universe:

Axiom B17

Galaxies, as well as clusters of galaxies, are considered to be point-like.

Axiom B18

These point-like galaxies constitute a continuous fluid.

In spite of such drastic simplifications, these hypotheses do not allow a complete determination of the matter/energy tensor: the number of independent components of this tensor is still too large to be unambiguously determined by such observations as are achievable from the solar system. The incorporation of new axiomatic propositions proves again to be necessary.

Axiom B19

The universe is isotropic about us.[37]

As far as possible and feasible,[38] it has been observationally verified that the overall distribution of galaxies and of clusters of galaxies is approximately similar in every direction of the sky. Of course, such isotropy can only be established for very large distance scales (about 100 Mpc; some 326 million light years), since in our neighborhood the universe presents a indisputable anisotropic ("clumpy") structure. This approximate isotropy has been strongly contested by several authors in recent years (1990-1991), but it remains one of the most essential assumptions of the standard Big Bang model.

Because this new hypothesis is not powerful enough to achieve our aim,[39] another assumption, called the "Cosmological Principle" or "Copernican Principle," states:

Axiom B20

We do not occupy a privileged position in the universe.

Consequently, the universe ought to be isotropic for *all* observers, anywhere. This idea can be extended to the concept of isotropy about a world line. It means that at every moment of an observer's history, all his observations are isotropic. Extrapolation to isotropy about every world line leads to the following postulate:

Axiom B21

The universe is homogeneous; it is spherically symmetric overall.[40]

If these postulates are admitted, the Einstein field equations can be radically simplified, and a metric function, specific to such a universe, can now be formulated. The result of all these simplifications is called the "Friedmann-Robertson-Walker (FRW)" model of the universe, after the Russian physicist A.A. Friedmann, who contributed fundamentally to this solution in 1922, and the equally fundamental contributions of the Americans H.P. Robertson and A.G. Walker in 1935, who gave the first proof that this model describes the most general Riemannian geometry compatible with a homogeneous and isotropic space-time. This FRW model is certainly not the only one possible. But, for the time being, it is favored by a majority of scientists. The resulting FRW metric function allows us to calculate the interval between two space-time point/events of the FRW universe; this interval, in spherical coordinates, is given by the formula

[equation 5]

$$ds^2 = -dt^2 + R^2(t)\{ dr^2 / 1 - kr^2 + r^2[d\theta^2 + sin^2\theta \, d\phi^2] \}$$

where r, θ, and ϕ are comoving spherical coordinates, which means that the point-like galaxies have no proper movement with respect to the frame of reference of the FRW universe: they move (comove) with the universe (and c is taken $= 1$).

$R(t)$ is an arbitrary scale factor, the "cosmic scale factor," having a specific meaning in this FRW model. The distance between comoving points in this kind of universe (galaxies and galaxy clusters) is scaled with $R(t)$, as is apparent from equation [5]. The determination of its value in terms of the matter/energy content of the universe gives the dynamics of the FRW model. However, as $R(t)$ changes with the matter density of the universe, the result, surprising at the time of its inception, is that the FRW universe is not a static universe but is instead a dynamic universe.[41]

Let us first look at the kind of geometry which now results. As seen in the above-given expression for the FRW metric, i.e., in equation [5], an arbitrary parameter k appears. This parameter is related to the curvature of the $t = constant$ slices. All such slices have, in this spherically symmetric FRW model, constant curvature. But the parameter k can take three possible values: for $k = 0$ the three-dimensional hypersurfaces defined by $t = constant$ are spatially exactly flat, and the FRW universe is open and infinite; for all $k < 0$ these slices are negatively curved, but similarly giving FRW universes that are spatially open and infinite in size; finally, for $k > 0$ the slices have a positive curvature resulting in spatially closed and finite FRW universes. Open universes are supposed to expand forever, whereas closed universes are believed to eventually enter a contracting phase; more details will be given shortly.

But the simplifying assumptions, so far propounded, are still not sufficient to decide the value of k, neither can they settle what kind of curvature the FRW universe has.

On the other hand, to determine the universal scale factor $R(t)$ of the FRW model it is first necessary to specify the matter/energy tensor T. Once again, new simplifying assumptions are postulated. We have mentioned above that galaxies are reduced to points and are supposed to constitute a continuous fluid. Now we add

Axiom B22

The galactic fluid behaves like a perfect fluid.

This assumption allows the study of the galactic fluid leaving aside the vexing problems posed by viscosity, shear forces, thermodynamic nonadiabatic conditions, and so forth.

Axioms B15 through B22 enable us to proceed to the final step: verification of the model by means of observational facts.

B) The Relation Between the Model and Observation

With the help of these new postulates, one can write down an equation of the state of the universe, relating the pressure p of this fluid

to its matter/energy density ρ. In this equation the two matter/energy
ingredients must be analyzed: the matter content of the "dust" of
galaxies, and the energy contribution of the radiation present in the
universe.

We now obtain the following expression:

[equation 6]

$$\dot{R}^2(t) - (8\pi G/3) \bullet \rho R^2(t) = -k ,$$

the dot over $R(t)$ indicates a derivation with respect to time t. The
universal scale factor $R(t)$ is not a constant value, as would be the case
in a static universe; on the contrary, $R(t)$ changes, as a function of the
value t of time: this universe is a dynamic object.[42] Furthermore, at
some particular value of t, the scale factor vanishes, $R(t) = 0$. This
represents a singularity of the manifold: the "Big Bang singularity."
This particular value of $t = t_{big\ bang}$ corresponds to a state in which the
three-dimensional slice defined by $t = t_{big\ bang}$ has volume equal to
zero, and in which the entire FRW universe is compressed into an
infinitely dense (and infinitely hot) point. For unknown reasons, the
FRW universe exploded from this "singular" state; therefore, this
cosmological model was given the name "Big Bang." As mentioned,
in a singularity, the laws of physics break down. Thus, the question of
what was "before" the Big Bang loses all meaning: time and space, as
dimensions of the universal manifold, come into existence with the Big
Bang.[43]

The actual value of $R(t)$ can be inferred only from observation.
Therefore, the question whether, at the present epoch, the universe is
expanding (open universe), or whether the universe is in a contracting
phase (closed universe), because the gravitational forces acting on the
exploding matter have overcome the expansion, can be settled by
observation only.

But what has observation revealed? Thus far, only three major
experimental facts connect the FRW model with observation: the
discovery of the Hubble law relating distance with redshift, the

detection of a cosmic microwave background radiation, and the theoretical analysis of the primordial nucleosynthesis of light elements predicting their presently observed cosmic abundance. These are the three most important observational facts supporting the isotropic and homogeneous FRW model of the universe; they strengthen the acceptance of this model by the majority of the scientific community.

The Hubble Law

The first detection of a radial velocity of a galaxy, Andromeda, approaching us at approximately 200 km/sec. was made in 1912 by the astrophysicist V. M. Slipher. A few years later, thirteen other galaxies were studied, all but two receding from us at roughly 300 km/sec. In 1929, the American astrophysicist E. P. Hubble announced he had found that the radiation received from 36 out of 41 far away galaxies - up to six million light-years away - was shifted towards the red end of the electromagnetic spectrum, showing properties that could be interpreted as a Doppler effect.[44] This "redshift" led the astrophysicist to propose what is now called the Hubble law: galaxies recede from our observational point in a roughly linear manner with distance; in other words, the universe is expanding. Needless to say, this discovery constitutes one of the key observational supports of the model.

The Hubble law establishes a rough direct-proportionality relation between the recession velocities of some galaxies, inferred from their respective redshift,[45] and their distance form the Earth, as deduced from independent data. The estimation of distances for far-away galaxies is based on a stepwise constructed distance ladder, utilizing for each rung a suitable distance indicator. Primary distance indicators are calibrated from observations in our galaxy (the most important: the Cepheid variable stars); secondary indicators depend on the knowledge of the distance to some representative nearby galaxies; and tertiary indicators rely on such features as spiral galaxy luminosity, the luminosity of the brightest galaxy in a cluster, and other global properties of galaxies.

Since, in the FRW model, galaxies are comoving with the $t =$

constant slices, so that they have no proper motion with respect to these slices, the Hubble law allows us to evaluate in a straightforward way the universal factor $R(t)$ or, what amounts to the same it, allows us to estimate the value of the Hubble "constant," which is not actually a constant value, because it is found by

[equation 8]

$$H(t) \equiv R(t)/R(t),$$

the dot over $R(t)$ again means derivation respect to time t. The value of this Hubble constant "today" $\equiv H_O$, is estimated to lie in the range between

50 $km.sec.^{-1}$ Mpc^{-1} (the Sandage-Tammann value) and
100 $km.sec.^{-1}$ Mpc^{-1} (the G. de Vaucouleurs value).

This rather substantial disagreement on the value of the Hubble constant should not surprise us: it is a consequence of differing appreciations of the calibration of the distance indicators.

The reciprocal of H is the "Hubble time," H^{-1}. This quantity represents the time the "point-galaxies" of the cosmological fluid would have taken to attain their present separation, if they had, starting from a condition of infinite compactness, always maintained their present velocities. This gives an approximate estimate of 10-20 billion years for the "age of the FRW universe." The fact that the universe is presently expanding does not settle the question of whether it is an open or a closed universe, whether such expansion will continue forever, until the universe vanishes into ever smaller densities, or whether a contracting phase will end in the inverse of the Big Bang, the "Big Crunch." Half a century after Hubble's influential discovery, the other two observational facts supporting the FRW model were established: the cosmic microwave radiation and cosmic abundance of light elements.

Cosmic Microwave Background Radiation

Cosmologists have established a detailed scenario describing the first minutes of the FRW universe following its explosive "creation." The different phases which the primordial universe went through have thus been described with a surprising degree of sophistication. If, in order to simplify their scenario, an important number of additional presuppositions is admitted, and if the latest refinements of elementary particle physics are called upon, specifically the Great Unified Theories (GUTs), it becomes possible for cosmologists to describe the evolution of the primordial FRW universe, starting 10^{-43} seconds after the Big Bang!

In the present physical universe four types of interactions, four types of fundamental forces of nature, are known:
• the gravitational interaction;
• the electromagnetic interaction;
• the weak interaction;
• the strong interaction.
According to the most recent physical theories (GUTs), the difference between these four forces disappears at extremely high energies, such as those reached in the primordial FRW universe; for these extreme values, supersymmetry prevails. Supersymmetry theories predict that at energies of 10^{19} GeV (1 GeV = 10^9 electron-volts) and at temperatures of the order of 10^{32} K, all four interactions are unified into one primordial, unique, force.[46] According to S. Hawking, at that time the universe may have looked like a foam of primordial small black holes, perpetually evaporating and reappearing.

As this primordial FRW universe expanded, it simultaneously cooled and its temperature started to diminish. In this way different "threshold" were reached beyond which the postulated supersymmetry of the four interactions was "spontaneously" broken,[47] and these interactions "decouple" one after the other. The first to decouple was the gravitational interaction, when the universe was barely 10^{-43} seconds old. This was followed by the decoupling of the other three forces, a process which was completed when the universe attained the age of 10^{-9} seconds.

Under these primordial conditions, a perfect equilibrium was established between matter and energy. In view of the high values of energy and temperature prevailing at that epoch, the FRW universe was in a state of permanent creation and annihilation of pairs of particles and antiparticles; at this stage, the distinction between matter and energy is meaningless. But as the temperature and energy of the universe tapered off as a result of its expansion, stable particles appeared, and conventional physics began to become applicable. However, the density of this primordial universe was still so high that energy, i.e., photons, were unable to escape: all were immediately captured by free electrons which, at these high temperatures, are not integrated into atoms. Under these conditions, the universe was "opaque;" no photons could escape. During this photon/matter equilibrium, radiative energy acquired the thermal character of a black-body, as described by Planck's law.[48]

According to the standard model, approximately 100,000 years after the Big Bang, the cooling process had come to the point where matter and radiation ceased to be in thermal equilibrium (decoupling): at that point, matter ceased to be ionized, the electrons joined atoms, the opacity dropped sharply, and radiation ceased to be in contact with matter. The consequence of this decoupling was the sudden "transparency" of the universe: photons – radiation – could now freely escape.

Since this free radiation did not further interact with matter, that is, with free electrons, nothing could change the thermal character which this radiation had acquired at the time of the matter/radiation decoupling. It was thus possible to determine its thermal spectrum, i.e., the curve showing how much energy the radiation had at various wavelengths. This calculation was done by George Gamow in 1946. Since decoupling, this radiant energy, that is to say, these photons, had followed the expansion of the FRW universe, and thus their wavelength had expanded in the same proportion. This led Gamow to predict the existence of a presently observable "background radiation" presenting the spectrum of a Planck black-body at approximately 5 K.

The accidental discovery in 1965 by Penzias and Wilson of a dilute electromagnetic radiation, reaching us from every direction of the sky, and presenting a black-body spectrum of about 3 K, constitutes an enormous success for the predictive power of the FRW model. Interstellar absorption studies have shown that this 3 K radiation can also be found in other parts of the Milky Way. The Earth, and even our entire Galaxy, seem to be plunged into a freezer with the thermic properties of a Planck black-body at a very low temperature: 2.735 ± 0.06 K, according to the latest measurements. This even bath of microwaves consists of electromagnetic radiation, with wavelengths in the millimeter range. Moreover, its isotropy is remarkable: measurements show this radiation to be isotropic up to about one part in 10,000. Since this radiation has propagated unhindered practically since the Big Bang itself, it has, so to speak, imprinted upon it information about the space-time geometry of the FRW universe.

A further prediction of the FRW model has been put forward. Similar to photons, the neutrinos, those elusive elementary particles, have also been subject to a decoupling process. At an epoch estimated to be even closer to the Big Bang, the universe becomes transparent to the neutrinos. This then allows us to predict that we should now be observing a neutrino background of a temperature of approximately 2 K; undoubtedly, the detection of such a neutrino background would be interpreted as the most remarkable verification of the FRW model.

The Relative Abundance of Light Elements

After the primordial FRW universe had attained photon transparency, and matter and energy had decoupled, the presently known chemical elements, the primordial gases made out of these elements and finally the more complex structures that eventually were to make up the galaxies, the galaxy clusters as well as the stars, the planets, etc., gradually appeared. Following this scenario, the present abundances of the lightest chemical elements put the strongest constraints on the model. In effect, the FRW model predicts two stages:

• a primordial nucleosynthesis of only the light elements, and

• a nucleosynthesis of the heavy elements, which takes place much later, in the interior of the stars. These heavier elements are reinjected into interstellar space through the supernova explosion mechanism.

The primordial nucleosynthesis stage has been calculated in great detail. Barely three minutes after the Big Bang, the nuclear reactor "Big Bang" stopped operating, not before it succeeded in transforming about 25% of the mass of the universe into helium, leaving almost 75% of it in the form of hydrogen, plus traces of the first light elements. Such trace elements had, according to these calculations, the following relative abundances in % by mass:

Deuterium	:	1 / 1000
Lithium	:	1 / 100,000,000
Helium 3	:	1 / 1000

It must be stressed again that there exist very strong constraints in these calculations, so that it is hard to arrive at different results without violating some fundamental postulate of the FRW model. Thus the experimental verification of these rates must be seen as a convincing support of this model. On the other hand, to measure the relative universal abundances of these light elements from our Earth-bound observation point is a demanding task.[49] The recently obtained results from the analysis of the radiation received from diverse sources has given an outcome that is in a surprisingly good agreement with these theoretical predictions,[50] thus providing one of the most solid observational confirmations of the model.

Observational Limits in Cosmology

In the preceding sections, we tried to provide a brief overview of the FRW model, as well as of the observational facts presumed to validate this bold theory. In order to further elucidate the epistemological structure underlying the model, we will now focus on the problem from a different perspective.

We wish to explore the question: what definite conclusions could possibly be gleaned from the sole accumulation of all obtainable astronomical observations? In other words: what is the minimum set of indispensable axiomatic propositions that must be admitted *a priori* in order to make such observations meaningful?

Although this subject will be discussed in detail in the third part of this book, we shall here begin by pointing out two distinct procedures that may be adopted by scientists.

1) As a first step, the scientist assumes as true[51] some formalized system of axiomatic propositions derived from generally admitted preconceptions or prejudices, and then tries to verify *a posteriori* by observation the solutions deduced within the framework of a (mathematical) theory incorporating these presuppositions as axioms; this was essentially the way we presented the standard model in the previous sections; or

2) The scientist takes as his exclusive starting point a set of observationally acquired data with the purpose of finding the most parsimonious and simple axiomatic system by which such data, then considered as solutions of the (mathematical) theory incorporating these presuppositions, can be deduced or "explained."[52]

It goes without saying that, in current scientific practice, both procedures are closely intertwined, which probably explains some frequently encountered misconceptions and confusions.

Restricting our inquiry to the cosmological problem, we will explore the second alternative, basing our analysis on the interesting results obtained by G.F.R. Ellis and his collaborators.[53] A more detailed and general analysis of this crucial procedure will be conducted in the third part.

Suppose the measurement of all obtainable (from the solar system) astronomical observational quantities can be conducted with unlimited accuracy and that, under these circumstances, one now asks: relying only on such "ideal" astronomical observations, what is and what is not decidable in cosmology? In other words, if one does not presuppose a theoretical model or, more precisely, if one does not take for granted that we know the dynamic laws determining the very large-scale structure of the universe, one asks: what can be inferred from these ideal observations about the space-time metric and the matter content of the universe? That is, what could be inferred if we were in possession of such accurate astronomical data, without taking *a priori* for granted that the Einstein field equations correctly describe the dynamics of the universe, and *without* assuming *a priori* that the universe is isotropic and homogeneous? In summary, we ask: what are the minimal assumptions to be added to the set of astronomical observations, in order to build a consistent cosmological model?

In the work of Ellis *et al.* that we have cited above the following minimum presuppositions are, in addition to "ideal" measurements, taken to hold true:

1) our locally valid physical laws hold everywhere in the universe, at all points of space-time;

2) the universe may be modeled with sufficient accuracy by a 4-dimensional manifold (plus its metric function);

3) the matter/energy content of the universe can be appropriately represented by the use of tensor calculus. In addition, it is assumed that at "suitably large" scales the galaxies and intergalactic matter can be approximated to a cosmological fluid, the galaxies reducing to points. Put differently, it is accepted as a postulate that the universe is enormously large.

Equipped with these basic hypotheses (cf. the complete list of geometrical and dynamical axiomatic propositions of the standard approach to cosmology), we now ask: what can be inferred from an ideal set of accurate astronomical observations?

The first insight follows from the enormous size of the universe. Human observations, ideal or otherwise, are restricted to only one *hic et nunc*, to only one "point" of one world-line, because we are unable to move off the world-line of our local group of galaxies. Taking this limitation into account, that is, our point-like observational basis in an enormous universe, observations that provide interesting cosmological information can be classified as follows:

a) *Local physical experiments.* These observations lead to local physical laws which certainly contain some interesting information for cosmology, considering the axiomatic extrapolation of their range to universal validity. Such observations (for example, experiments to verify Einstein's gravitation, or experiments about the electromagnetic spectrum) are, moreover, repeatable.

b) *Local geophysical and astrophysical observations.* These observations provide valuable information about our present situation and about the history of our corner of the universe. Without additional assumptions, this information does not, however, tell us anything about distant regions of the universe. Nevertheless, several methods for estimating the age of the Earth, of the Solar System, and of nearby globular clusters of stars for example, set constraints for the age of the universe.

c) *Extragalactic high-energy massive particles.* Cosmic rays do convey useful data, but do not directly give detailed information about distant regions: one has first to establish that they actually are extra-galactic, and then to relate them to specific sources; both tasks are difficult.

d) *Astronomical observations of distant objects and of background radiation.* These observations can be classified as "cosmological information" properly speaking, provided all the assumptions about the nature and the propagation of electromagnetic energy are adopted. In the last decades, such radiation became measurable throughout its entire spectrum: from the hardest gamma-rays, through X-rays, ultra-violet, visible, infra-red, to microwave and radio frequencies. However, we only receive signals originating within our past light cone. All the "elsewhere" of the universe stays absolutely out of reach:

Inventing the Universe

this strongly constrains our possibility of building a model of the entire universe based on this limited source of information.

Ellis *et al.* thus arrive at the following conclusion: the analysis of the ideal observational data can provide us with rather little information about the universe, even if we restrict our ambitions to understanding the space-time structure only in the vicinity of our past light cone. In particular, we are unable to determine by direct observations the relationship between the radial coordinate down our light cone and the distance, as measured for instance by the redshift z, to any object we observe.[54] Furthermore, as an example of this observational indeterminacy, we cannot prove, based only on ideal observations, whether or not space-time is spherically symmetric about our position – even if all observations are isotropic around us. Even more striking: if we take for granted the universe has a FRW geometry, we still cannot, relying on "ideal" observations, say whether spatial slices (as given by the value of k in the FRW metric, cf. equation [5]) have zero, negative, or positive curvature. Nor do isotropic observations, together with the assumption of spherical symmetry, prove that the universe has a FRW geometry.

It goes without saying that the material limits of what can be observed impose even more severe limits on what we can hope to prove directly from observations. Thus, relying exclusively on observations, and taking for granted minimal assumptions, it remains impossible in principle to determine major features of the space-time structure of our past light cone – not to mention what lies outside of this light cone.

Finally, Ellis *et al.* look into the question of what happens if the whole body of postulates underlying the Einstein field equations[55] is admitted. Here the results are more encouraging. Under these enlarged (more "complex") hypotheses a spherically symmetric pressure-free (i.e., galactic "dust") model is spatially homogeneous, if and only if precisely the FRW relations obtain in this universe. And surprisingly the "ideal" cosmological observational data are exactly necessary and sufficient to determine the space-time structure of our past light cone.

There is no *a priori* reason why this should hold true. Space-time could have been underdetermined by such observations, resulting in the theoretical impossibility of ascertaining the geometry of our past light cone by astronomical observations, no matter how complete or accurate. This was the state of affairs before the introduction of Einstein's field equations as an axiom. Conversely, space-time could have been overdetermined, so that consistency conditions would have to be fulfilled between these ideal observations, if Einstein's field equations were to be satisfied. Neither is the case, and it has thus been proved that (ideal) astronomic observational data can in fact be used to confirm the FRW model.

Although real observations differ altogether from ideal ones, analysis of an ideal model nonetheless provides a measure of the indispensable minimum of *a priori* propositions,[56] without which no meaning can be assigned to observationally collected information, even in cases where perfectly error-free data is available.

What conclusions can be drawn from these results? It has been repeatedly emphasized that sense-perception data cannot lead to any intelligible explanation independently from a formalized axiomatic theory of some kind. In everyday, local, earth-bound experience, such a theory is a prerequisite for any meaningful interpretation of observational information. In cosmology the situation is more complicated. The object considered is our unique, irreplaceable and unrepeatable universe. Moreover, we have no access to any event outside the horizon of our past light cone,[57] which therefore constitutes our only source of information. That is why the building of a cosmological model requires, for Plato as well as for contemporary scientists, bold extrapolations, introduced as drastic simplifications in the analysis of the available observations (homogeneity, galaxies as points, and so forth).

From these considerations it follows naturally that the FRW model is beset by very profound difficulties. Actually, such aberrations as have been brought to light falsify the model altogether. However, as is well documented, models are seldom (if ever) abandoned before a new and clearly better one is introduced.[58] Lacking, for the time being, any such "fairer" alternative, scientific methodology consists in proposing additional *ad hoc* postulates, without changing the fundamental set of *a priori* propositions on which the model is based ("adding epicycles").

Problems Affecting the FRW Model;
Observations Requiring Additional Axioms

The standard FRW model, based on Einstein's equations and the FRW symmetry (homogeneity and isotropy), explains the expansion of the universe, the presence of a cosmic microwave background radiation, and predicts with accuracy the observed light element abundances. It further seems to give an adequate accounting of the evolution of the universe starting from 0.01 sec. after the Big Bang up to the present day (some 10-20 billion years later). But it is not devoid of severe problems.

The Horizon Problem

The suppositions underpinning the FRW model assume a homogeneous space-time geometry. Such assumptions have been supported by the discovery of the high degree of isotropy – of the order of one part in 10,000 – of the microwave background radiation. However, it is precisely this isotropy that lies at the heart of the horizon problem.

The "horizon size" given by *1/H*, where *H* is the Hubble constant for some epoch at time *t*, calculated in equation [8] as a function of *R(t)*, the scale factor which determines the intergalactic distances, is the distance light can travel from the beginning of the cosmological expansion to a specified time *t*. This distance, in turn, fixes the limiting distance, "the horizon," inside which causal interactions can take place. But if some region of the universe lies outside the horizon of some other region, at some time *t* after the Big Bang, then these regions are causally disconnected at time *t*: they have not yet had time to communicate with each other via light signals.

It can be deduced from the FRW model that matter in an expanding universe decelerates due to gravitational attraction. Thus, the ratio of the distance between objects at fixed coordinate locations to the horizon size also decreases with time. That is, the horizon grows more rapidly than the expansion rate between two objects. As a result, if two objects, or two regions of the universe, fall outside each other's horizon today, that is, if they are separated by a distance greater than *1/H* today, then they have always been outside each other's horizon.

To help provide a better understanding of this important line of argumentation, imagine the universe as a sphere, so that one can symbolize, as in Figure 2.15, the universe at three different moments of its expansion:

Figure 2.15

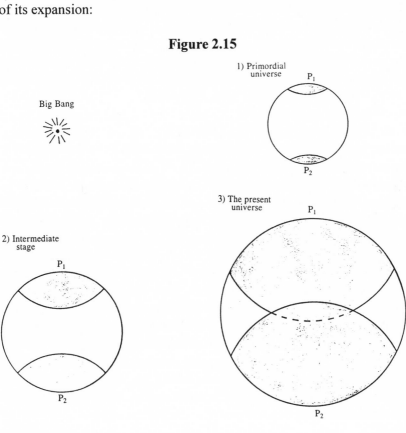

At each "pole" of this spherical universe we indicate two events, P_1 and P_2, with their respective horizons at each epoch indicated by shading. The horizon of a given event is the distance from beyond which light signals would not yet have had time to reach a given event/point. In the very early (radiation dominated) universe, the state equations of the FRW model show that the "radius" of the universe grows like the square root of the time t, whereas the distance of each "pole" to its horizon grows in direct proportion to the time t. In consequence, the further we go back in time, the horizon encloses an ever smaller portion of the universe. It follows from Figure 2.15 that the events P_1 and P_2 can both be "seen" by one (Earth-bound) observer today only if he is situated where the shaded parts overlap. On the other hand, since they are at present still outside of each other's horizon, the events P_1 and P_2 could never have interacted causally in the past. They can do so only at some future time when one of them falls inside the horizon of the other: only at that future time, one event "sees" – can causally interact with – the other.

When we measure the microwave background radiation coming from two regions of the sky separated by 180°, what are we really measuring? In the FRW model, such radiation was emitted when the universe was only 100,000 years old. Consequently, since this emission has only just reached us today, we are clearly looking at regions which have not yet come into each other's horizon; this will happen in about 10 billion years. It follows that these two regions have never been in causal contact. Yet their temperatures are the same, up to present experimental accuracy. If this is the case, how did regions which were never in causal contact come to have essentially exactly the same temperature? What is more, it can be shown that the early universe consisted of some 10^{83} causally disconnected fragments: what kind of "fine-tuning" gave them all the same temperature? Unless one assumes that the forces which created these initial conditions where capable of violating the causality axioms of Einstein's relativity, one must accept this striking fine-tuning as some kind of initial condition of the Big Bang itself, or equivalently, admit 10^{83} additional axioms (epicycles)!

The Flatness Problem

According to equation [5], the FRW metric includes a parameter k, and the curvature of the spatial slices of the universe depends on whether this parameter is negative, positive, or zero. By rescaling the value of $R(t)$ in the FRW metric, one can always make $k = -1$ (*if* $k<0$), and similarly $k = 1$ (*if* $k > 0$); $k = 0$ is a limit of one of these two cases. How can the value of k be deduced from observation?

If critical density is defined as the matter/energy density of the universe[59] resulting in $k = 0$, and if the ratio of this critical density to the density of the universe today is defined as Ω, then one can prove that in the FRW universe

[equation 9]

$$\Omega = 1 + k / \dot{R}(t)^2,$$

where the dot over $R(t)$ stands for time derivation. The time derivative of the universal scale factor $R(t)$ decreases with time. Since gravitational attraction decelerates the expansion following the Big Bang explosion, the growth of the scale factor decreases; whether it will reverse its expansion into a Big Crunch or continue forever overcoming gravitational attraction, depends on k, that is, on Ω.

According to equation [9], after the Big Bang the value of Ω (for $k = \pm 1$) deviates further and further from unity, and only for k exactly equal 0 will Ω remain constant, that is, will Ω remain equal to unity. For any value of Ω slightly different from unity, and for $k = \pm 1$, the universe would have long ago either recontracted into the Big Crunch, or flown away into the most tenuous densities. But, from observationally gathered data it has been deduced that the present-day value of Ω falls somewhere in the range between 0.01 and 10. This means that the present-day value of Ω is very close to unity and, furthermore, that Ω has remained this remarkably near to unity for the about 10 or 20 billion years of expansion of the universe.[60] Again, this fact requires a surprisingly precise fine-tuning of the Big Bang: the initial value of Ω must have been equal to unity to incredible precision.

How did this happen? This is the flatness problem, so-called because for $\Omega = 1$, the universe is exactly flat.

The Dark Matter, or Missing Matter, Problem

Matter in the universe is matter we see, that is, matter we detect by its emission (or absorption) of electromagnetic radiation. Thus we "see" stars, galaxies, pulsars, quasars, intergalactic gas, and so forth. Even black holes can be "seen" by indirect evidence. All such matter is given the name of "luminous matter." Such matter always interacts with electromagnetic radiation in some way. But according the FRW model, this type of matter represents only a minimal fraction of the matter content of the universe, perhaps less than 1% .

In the first place, it has been known for some time that galaxies must contain more matter than just the luminous matter. Observations of spiral and elliptical galaxies can determine the rotational velocity of such a galaxy. Galaxies are supposed to be long-lived systems, their life-span being estimated to be of the order of 10^9 years. What holds such a long-lived rotating system together, and stops it from flying apart due to its kinetic energy? As far as we know, only gravitation. In that case however, the rotational velocities must depend in a well-defined manner on the radius r taken from the center of the galaxy where, in general, most of the luminous matter is concentrated. However, aberrant behavior[61] is consistently observed. Only one explanation seems to remove this problem: that the halos of galaxies must contain enormous quantities of dark matter, called "dark" precisely because it is neither seen, nor has any interactions of this matter with electromagnetic radiation so far been detected.

A further point is even more important. In relation with the flatness problem, we mentioned above that the value of Ω must be (almost) exactly unity (representing a critical density of about 12 atoms per cubic meter). However, the luminous matter accounts, according the latest estimates, only for a value of Ω of the order of 0.007 (equivalent to approximately 0.1 atoms per cubic meter). Dark matter in halos of galaxies cannot account for Ω larger than about 0.4. Another factor must also be taken into consideration. Primordial nucleosynthesis is the strongest numerical validation of the FRW model. However, the nucleosynthesis of light elements imposes strong constraints on the amount of matter that can be present in the universe in baryonic form. Baryonic matter is ordinary matter, protons and neutrons, out of which matter, such as it is known in the solar system, is made of. The maximum value of Ω is of the order of 0.14, if the matter in the universe is exclusively in baryonic form, compatible with the successful predictions of the relative abundances of light elements and with the theory of primordial nucleosynthesis.

The upshot of all this is that, in order to maintain the consistency of the FRW model with general physics, one must postulate the universe to be made of anywhere between 80% and 99% of some exotic, non-baryonic, weakly interacting (weak enough to escape detection so far), dark matter.[62] Such exotic forms of matter have already been conjectured in some of the Great Unified Theories. Furthermore, due to other types of considerations, this dark matter is most probably not clustered together with the luminous matter, its universal distribution is different (and unknown).

The dark matter problem becomes more acute when one adds to the FRW model the inflationary scenario. This scenario's strongest prediction is precisely that $\Omega = 1$, making the eventual detection of this exotic but ubiquitous form of matter one of the preeminent aspirations of contemporary cosmology.

The Inflationary Scenario

The Grand Unified Theories (GUTs) predict that at very high energies, the four known interactions: gravitation, electromagnetism, the weak and the strong nuclear interactions, can all be explained by one unique, and hopefully simple, theory. At such high energies, which are unattainable in terrestrial laboratories, there would be just one type of symmetry group describing all matter. If the FRW model is the correct description of the universe, then immediately after the Big Bang the universe would have found itself precisely in this unifying energy range. Present day matter and its four interactions are deemed to be the consequence of a spontaneous symmetry-breaking process, setting in shortly after the primeval explosion, and having random quantum fluctuations as its cause. It thus happened that the primordial universe was adopted by the GUT-theorists as their ultimate "laboratory." And cosmology and particle physics merged into a common research space; from this joint effort came inflation.[63] Bringing these disparate research fields together is tantamount to incorporating their mutually independent presuppositions into the list of axioms underpinning the FRW model of the universe.

Obviously, we cannot hope to summarize the inflationary scenario in this book. Suffice it to say that this model predicts that the universe, at about 10^{-35} seconds after the Big Bang, underwent a phenomenal expansion in an extremely short lapse of time. Between 10^{-35} seconds and 10^{-33} secs. after the Big Bang, one can construct a scenario, based on quantum mechanical speculations, where this exponential inflation can be theoretically justified. Actually, all that is required for inflation to become feasible is that, during this short period, the expression

[equation 10]

$$\rho + 3p/c^2 < 0 \quad , \text{i.e., is negative,}$$

where ρ is the density of matter, p is the pressure of matter, and c is the speed of light. This means that in such an epoch the normally attractive force of gravity becomes repulsive, resulting in an explosive expansion of the universe.

This inflationary scenario predicts that the present density of the universe is such that Ω is almost exactly equal 1, i.e., that this density is at most one part in 10,000 away from the critical density. Thus, it must also predict that the universe is predominantly made from some exotic form of dark matter. The discovery of this dark matter, so far not accomplished, remains nevertheless the pivotal test of the model.

Why, despite its abstruse prediction of exotic dark matter, is the inflationary scenario the preferred model nowadays? Mainly because horizon and the flatness problems. During inflation the horizon grows so fast that it encompasses all that part of the early universe that will become "our" universe after the inflationary period. This means that points that are outside the horizon today were inside a single horizon – and thus causally connected – at some time in the past of the universe. This will then "explain" the observed isotropy of the microwave background radiation. But the price to pay is high: the explosive inflation does not produce just "one" universe. In fact, inflation somehow creates a conglomerate of "bubbles" and each of these gives rise to one separate universe.

In the *Timaeus*, to the question: "Are we right, then, in describing the universe as one, or would it be more correct to speak of heavens [universes] as many or infinite in number?" Timaeus answers: "One it must be" (*Timaeus* 31a). After inflation, some cosmologists reply differently. They invoke an "Anthropic Principle" (in the "weak" version) in order to answer the question: which of the many universes is ours? by saying: precisely the one whose physical constants (gravitational constant, speed of light, and many others) and other conditions (such as size, age, etc.) can make feasible the presence in the universe of rational observers such as we are. Thus, paraphrasing the *Timaeus*, their response to this question is: "Precisely this particular one it must be."[64]

Here ends our brief description of the standard FRW cosmology. Our goal was to parallel this model with Plato's cosmology in the *Timaeus*. To try to make this parallel complete, we conclude with a few glimpses at the contemporary theory of elementary particles, where symmetry plays a crucial role, as it was also the case in the *Timaeus*.

C) *Modern Theory of Matter*

Reducing the "visible and tangible" world and all the phenomena perceived by the senses to just a few, simple, invisible and intangible primordial components and to their mutual interactions, is an old dream, a dream that still remains alive today.

Modern physics postulates all matter to be made out of a reduced number of mathematically well characterized "elementary particles." Such particles have precise properties, and are considered to be simple, that is, no internal structure originating from an association of even more elementary particles can be discerned. They are supposed to interact exclusively through only four different types of forces. This means that all physical phenomena whatsoever can be reduced to the interplay of only these four types of interactions among the elementary particles, representing an enormous simplification and parsimony of description. These fundamental interactions are shown in Table 2.5:

Table 2.5

FORCE	RANGE	STRENGTH*	ACTING ON
Gravity	Infinite	6×10^{-39}	All particles
Weak Nuclear Force	$<10^{-18}$ m	1×10^{-5}	Leptons, Hadrons
Electromagnetism	Infinite	$1/137$	All charged particles
Strong Nuclear Force	10^{-15}	1	Hadrons

* Dimensionless, intrinsic strength.

Notes.
- 10^{-13}m is approximately the dimension of the atomic nucleus.
- Hadrons: the elementary particles such as the proton, the neutron, and other heavy particles. Modern theory assumes that these

particles have a deeper internal structure and are made up of even more elementary particles: the Quarks.

- Leptons: the elementary light particles: electron, muon, neutrino.
- Weak force: causes the radioactive decay of some atomic nuclei.
- Strong force: holds the atomic nucleus together.

Particles, subject to these four interactions, are in turn each characterized by a well defined set of quantum values describing their intrinsic properties, e.g.,: charge in units of the charge of the electron e, parity P, isospin I, baryon number B, lepton number l, spin J in units of the Planck constant h divided by 2π, mass in MeV,[65] average lifetime τ in seconds, the main modes of decay of one particle into one - or many - other(s), etc.

Hundreds of such fleetingly stable forms of elementary matter are known. Fortunately, they can be classified into a restricted number of "families" a brief description follows.

1) *Leptons* (from the Greek *leptos*: fine, light-weight) are spin = ½ particles which include the electron and its corresponding antimatter particle (antiparticle) the positron, the muons and the neutrinos, all with their corresponding antiparticles. Leptons do not "feel" the strong force. Initially their masses were believed to be less than about 100 MeV.

2) *Hadrons* (from the Greek *hadros*: large, heavy) include the *mesons* (from the Greek *meson*: between), particles with integer spin and masses initially believed to be between those of the leptons and the baryons, and the *baryons* (from the Greek *barys*: heavy, weighty), massive half integer spin particles subdivided in turn into *nucleons*, i.e., the proton and the neutron, and the "strange" *hyperons* and *kaons*. Hadrons are distinguished by the fact that they feel the strong interaction: this is the force which holds neutrons and protons together within the nucleus.[66] Ordinary matter, out of which the tangible and visible universe is made of, includes nucleons, electrons, and neutrinos (plus photons, i.e., radiation); all the other particles are unstable in the sense that they decay very fast into these more stable particles.

However, the free neutron is also unstable – its lifetime is about 925 seconds[67] – and even the proton is predicted to be unstable, with lifetime of the order of 10^{31} - 10^{32} years; experiments have not yet detected the decay of this fundamental particle.[68]

3) The last family is constituted by particles called *field quanta*. The photon is the quanta of the electromagnetic field, for the other interactions, different field quanta are defined; the gravitational field quanta, for instance, is the graviton. Field quanta are all bosons (see below); after their theoretic prediction, most of them have been experimentally found in particle accelerators. A recent theory has succeeded in "unifying" the electromagnetic interaction and the weak interaction: one unique theory rather adequately describes the "electroweak" field. Among its field quanta are the "intermediate vector bosons" their discovery in 1983 at the Centre Européen de la Recherche Nucléaire in Geneva (CERN) represents a stupendous success of this theory.

A further important subdivision of the elementary particles is into classes called *bosons* and *fermions*, according to whether their spin has an integer or a half integer value in units of Planck's constant h divided by 2π. Fermions obey the Pauli exclusion principle: no two fermions are allowed to share exactly the same quantum state. This property shows up in the kind of statistics or probability predictions that can be established, in the sense that it reflects the possibility of distinguishing one particle from another. The Pauli principle has wide-reaching consequences. Electrons are fermions, their spin is ½. In an atom, where electrons occupy shells according to energy (and other parameters), no two electrons can occupy exactly the same place on a shell. As shells have only limited available places, electron occupancy of shells follows a strict order. This intra-atomic ordering scheme leads precisely to Mendeleyev's periodic table of the chemical elements. It follows that chemistry, in the Mendeleyev picture, *reduces* to quantum mechanics, as postulated. On the other hand, photons are bosons, their spin is 1. They are not subject to Pauli's exclusion principle and thus can all occupy the same energy level. Since photons can share the same energy state, they can be made to travel in coherent light beams as in a laser.

The Central Role of the Concept of Symmetry

The concept of symmetry which intervenes, at different levels (isotropy, homogeneity, perfect galaxy gas, GUTs, supersymmetry, symmetry breaking etc.), in the building of the mathematical theories modeling the universe, plays also an important role in modern particle physics; four classes of such symmetries will now briefly be analyzed.

1) Continuous space-time symmetries

The rigorous study of the notion of symmetry is the subject matter of a branch of mathematics called "group theory." Broadly speaking, a group is a collection of elements with specific interrelationships defined by group transformations. This notion is made non-trivial by the demand that repeated transformations between elements of a group should be equivalent to another group transformation, that is, equivalent to a single transformation from the initial to the final element.

Foremost amongst the continuous space-time symmetries are the symmetries having as their consequence the invariance of the physical phenomena under the operations of translation through space, translation through time and rotation about an axis. But the laws of physics should be so formulated that any physical phenomenon remains independent of the choice of a particular frame of reference (system of coordinates). This entails that in such a description the mathematical expression of the physical laws should be invariant under all these transformations.

When a symmetry group governs a particular physical phenomenon, this condition requires that the mathematical expression of the dynamics of this phenomenon, in this case the Lagrangian function[69] governing it, remains invariant under the group transformations. A proof given in 1918 by Emmy Noether of a fundamental mathematical theorem ("Noether's theorem") states that for every continuous symmetry of a Lagrangian, there is a quantity conserved by its dynamics.[70]

The application of Noether's theorem then shows that invariance under a translation in space implies the conservation of momentum, that invariance under translation in time implies the law of the conservation of energy, and that invariance under spatial rotations implies the conservation of angular momentum. Thus, *the most fundamental conservation laws of physics are seen as consequences of the existence of symmetries.*

In practice, these three symmetries are duly incorporated into the description of the FRW universe if it is specified that the system is invariant under the group of "Lorentz" transformations. In fact, in the model generating the FRW description of the universe one can always, locally, construct a Lorentz invariant frame of reference in which the fundamental physical laws of conservation hold.

2) Discrete symmetries

The continuous symmetries of the preceding paragraphs derive their name from the fact that they can be built up from a succession of infinitesimally small steps. By contrast, discrete particle physics symmetries cannot be so constructed, and their associated conservation laws do not entail such a sweeping importance. But they determine in many instances the allowed behavior of some particle reaction. For the sake of completeness, we shall briefly mention:

Parity inversion (P), the operation under which a system is reflected through the origin of coordinates or, equivalently, when a right-handed coordinate system is reversed into a left-handed one. It is believed that electromagnetic force conserves parity, meaning that electromagnetic transitions can only occur between states of the same parity *P*. This limits the allowed transitions and in this way prescribes certain energy values to the photons emitted (or absorbed) in such a process, which in turn can be experimentally tested. Similarly, the strong interactions are invariant under the *P* transformation, whereas the weak interaction might violate this symmetry.

- *Charge conjugation (C)*, means that the laws of physics that can be established for a set of particles remain invariant for the same set of antiparticles. Only the strong and electromagnetic interactions are strictly invariant under the *C* transformation.

- *Time reversal (T)*, refers to the operation which changes the direction of time (the arrow of time) of some process. It states that a system is invariant under time reversal, if its evolution from an initial state to some final state can be reversed by changing the direction of time: it will then evolve from that final state to precisely the same initial state.

- *CPT-invariance*. By operating all these three symmetries simultaneously, their product symmetry is obtained. Experiments have shown that in nature the *C* and the *P* and the product *CP* symmetries are in certain cases violated. This means that one can devise experiments in which nature differentiates between left or right (mirror images), or between particle and antiparticle. These deviations came as quite a surprise to the physics community. However, it is generally believed that the *CPT* symmetry is absolutely exact and cannot be violated under any circumstances.

3) Dynamic symmetries

One example is the conservation of electrical charge. These symmetries are formalized as shifts in phase and other manipulations of the Lagrangian describing the system. These invariances give rise to significant predictions, such as the existence of new particles and their corresponding electric charge, spin, and other quantum numbers.

4) Internal symmetries

Particle physics was led by experiment to assign ever new quantum properties to the elementary components in order to find a representation able to describe the results of the interactions observed in particle accelerators. They were given whimsical names such as Strangeness, Charm, Color (blue, red, green, magenta, etc.), Flavor, and so on. Suffice it to say that the symmetry invariance approach is the essential tool which has allowed scientists to establish a classification of the ever-growing "zoo" of these presumed elementary particles.

The importance of this point cannot be over-emphasized: symmetries are the most fundamental "explanation" for the way things behave (the "laws" of physics) in the universe. Symmetry manifests itself in that the basic mathematical description of some physical systems does not change under certain transformations. Thus, symmetry is the fundamental attribute conferring some degree of stability upon an ever-changing material world, and is consequently the feature which makes possible a scheme of classifications (taxonomy) of the physical world. Symmetry is that property which allows us to give names to things, to perform the measuring operation, and to allow comparisons to be established. Accordingly, for Plato the contrary to *summetros* is *ametros*, that disordered (complex) state where no measure is feasible, no comparison possible.[71]

Symmetry in the modern theory of matter is treated mathematically by group theory, as already mentioned. Let us consider the rotation group as an example.

The set R of the rotations of a system form a group, each possible rotation being a member of the group. Two successive rotations: R_1 followed by R_2, written as (the "product") R_1R_2, are equivalent to a single rotation, that is, to another group element. There is an identity element: "no rotation," and every rotation has an inverse: "rotate back." The product is not necessarily commutative, since it is possible that $R_1R_2 \neq R_2R_1$, but the associative law always holds: $R_3(R_2R_1) =$

$(R_3R_2)R_1$. The rotation group is a continuous group in that each rotation can be labeled by a set of continuously varying parameters. Finally, the rotation group being a Lie group,[72] any rotation can be expressed as the product of a succession of infinitesimal rotations.[73]

Physics should be independent of the particular frame of reference chosen for observation (viz. the laboratory walls). Therefore, the equations expressing the dynamic behavior of a system (of particles) must be written in an invariant form, that is, they must remain unchanged under the symmetry operation R. More precisely, the probability that a system described by some set of parameters be found in some physical state must be unchanged by R.

In elementary particle physics, in the particular case of infinitesimal rotations, one finds a quantum mechanical operator J_3 called the generator of rotations about the 3-axis (or z-axis). This operator has the property of being a "hermitian" operator, and hence corresponds to a quantum mechanical observable.[74] Further analysis shows that one can identify the generator J_3 of rotations about the 3- (or z-) axis, with the third-component of the angular momentum operator of the system. The "eigenvalues"[75] of the observable J_3 are *constants* of the motion of the system if the equation of motion of the system is to remain unchanged by the symmetry operation.

Thus, a symmetry of the system has led to a quantum mechanical conservation law, the conservation of angular momentum. Experiments performed in different orientations of some measuring apparatus having given the same results, it follows that the system has rotational symmetry, and the law of conservation of angular momentum ensues.

Unfortunately for our exposition, quantum mechanics requires highly sophisticated mathematics. Therefore we drastically abbreviate.

Physical phenomena in the quantum domain are most often portrayed by symmetry groups, and the key conservation laws of nature are shown to be a consequence of them (as requested by Noether's theorem). In connection with the symmetry properties exhibited by the spectra and reaction patterns of hadrons, a group-theoretical approach has been developed. An important example, to which we will return below, arises from the similarity in behavior of neutrons and protons in the nucleus of atoms, apart from their difference in electromagnetic properties (positive charge for the proton, no charge for the neutron). This similarity was formulated as the invariance (conservation) of a new quantum number: "isospin," and the associated symmetry is the isospin invariance under the group *SU(2)* (see below). In several other cases, symmetry properties are understood in terms of new additive quantum numbers that are conserved, or nearly conserved.

We have just mentioned the rotation group *R*. In quantum mechanics, the lowest-dimension non-trivial representation of the rotation group is described by the 2×2 Pauli matrices, and represents a spin-½ particle, with spin projections along the 3-axis either up (+ ½) or down (- ½). The set of all 2x2 matrices is known as the group *U(2)*. A special unitary subgroup of *U(2)* is a group in two dimension, the *SU(2)* group. This group applies in the case of nucleons, that is, protons and neutrons, and provides us with an illustrative example of the application of the symmetry concept in the classification of elementary particles.

If we leave aside their electric charge, the proton and the neutron are almost identical particles differing only slightly[76] in mass. Since the strong interaction totally ignores the effects of electric charge, from the standpoint of this interaction the proton and the neutron may be regarded as two states of the same particle: the nucleon. But they do have a different electric charge. Thus we must consider the nucleon as a particle having an internal degree of freedom with two allowed states: the proton and the neutron. Mathematically, this means that there exists a *SU(2)* symmetry group such that the neutron and the proton are the fundamental representation of this symmetry group. As with the Pauli matrices representing spin-½, here similar matrices are introduced representing the new quantum number *isospin*, designated *I*. The 3-axis projection of the isospin is called I_3. The two possible "rotations in isospin space" give two symmetric base states (neutron/proton) for the nucleon; they are said to form a "doublet." And the mass difference between proton and neutron is "explained" as the effect of electromagnetism on the two different base states of the nucleon.

Higher order symmetries are described by the 3x3 matrices of the *SU(3)* group. The fundamental representation of *SU(3)* is a "triplet," with three different base states. The numerous hadrons detected in particle accelerators led to the discovery of another additive quantum number: *strangeness*, designated by *S*. This suggested the application of the isospin/nucleon symmetry approach. By enlarging the new symmetries under isospin and strangeness with *SU(3)* the result was a "multiplet" representation. However, the *SU(3)* multiplet structure derived from linking the extra symmetry of strange and non-strange particles is a much more approximate one.[77]

The *SU(3)* multiplet structure of classification of the elementary particles reminds us of Mendeleyev's periodic classifications of the chemical elements, which led to the more fundamental underlying proton-neutron-electron structure of the atoms, in that this new classification also led to the assumption of even more elementary particles: the quarks. Once again, in physics, it was symmetry that revealed the most fundamental structures.

According the quark model, all hadrons are made up of a small variety of still more primordial entities, called quarks, bound together in different ways. As indicated, the fundamental representation of *SU(3)* is a triplet. From the basic quark triplet, all higher order multiplets can be built. This provides the essential mechanism for constructing a taxonomic structure in which the hadrons and the mesons, that is the particles feeling the strong force, can be classified. But this classification allows us not only to predict the behavior under interactions of the hadrons (baryons and mesons), but it can be used to predict the existence of new particles, such as the Ω^- , identified by an arrow in Figure 2.16. Its discovery in 1963 at the Brookhaven National Accelerator Laboratory (USA) was a striking confirmation of this approach.

Each quark is assigned a spin -½, a baryon number $B = 1/3$, and an electric charge in multiples of 1/3 the charge e of the electron. The model further postulates that all baryons are made up of three quarks, and all mesons of a quark-antiquark pair. For classification purposes a new additive quantum number, *hypercharge*, written Y, is defined as

$$Y \equiv B + S,$$

where S is strangeness and B is baryonic number.

This scheme allows a simple graphic representation of the multiplets. As an illustration, Figure 2.16 shows two baryon multiplets: the spin-½ octet and the spin-$^3/_2$ decuplet. The Greek capital letters in these diagrams represent various particles predicted and/or found to occur in particle accelerator experiments. We show these diagrams without further comment, except by stressing their evident symmetry.

This ends our much too compendious description of the mathematical model of contemporary Big Bang cosmology, the most advanced and most successful representation of what modern physics has to say about the universe. It cannot be disclaimed that a tremendous progress has been achieved, nor can we forget that equally substantial problems remain to be solved.

Up to this point, our purpose in this book was to sketch the axiomatic structure of two cosmological models: the model put forward by Plato in the *Timaeus*, and the contemporary Big Bang cosmological model, but certainly not to pass judgment upon either of them. As we proceeded, moreover, we tried to constantly keep in mind these fundamental aspects:

• No "scientific" *knowledge* is possible independent of some formalized axiomatic framework.
• Such a framework requires the previous precise and explicit listing of the admitted axioms.
• Based on these axioms, theorems can be mathematically or

Figure 2.16

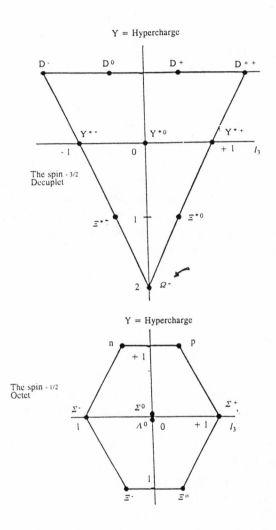

logically deduced; mathematics and logic are presumed to be clearly defined and well known.

• Only these theorems allow a connection with reality to be established. This means that such theorems must be capable of expressing the prediction of the results of some measurement to be performed on some physical system (for instance, the position of some heavenly body).

• The degree of agreement between prediction and observation furnishes a method, called the *scientific method*, which allows us to select the most adequate system of axioms.

• How these axioms are established by the scientist in the first place, whether by means of pure reasoning, aesthetics, or philosophical beliefs (Plato, Einstein), or by "inspired guesses" after examining some record of empirical measurements (Kepler), is inconsequential. Only the method, and the fact that "it works" are of consequence: axioms are established, theorems are deduced, predictions are made, and observations are used to sort out those axioms which are the most appropriate.

But the method necessarily must make use of mathematics and logic. They specify the rules of inference of the theorems, starting from the axioms. These rules are, however, nothing but a *mechanical* way of producing the theorems from the axioms, and they apply similarly in all models. Consequently, *the knowledge conveyed by this scientific method reduces to the enunciation of the* a priori *established list of axioms*.

Two further remarks are apposite here. 1) It is at the very least surprising that the list of axioms of the model proposed by Plato in the *Timaeus* in the fourth century BC., and the list of axioms underlying the contemporary Big Bang model, rely on the same abstract concepts. In effect, such notions as order, harmony, symmetry, measurement, stability, invariance, as well as the eternity and universality of the laws of nature, constitute the pivotal notions in both models. 2) The position of the axioms is enigmatic in both models. Whereas Plato appeals to intelligible Forms under the dominance of the Good plus a demiurge, which finally explains strictly nothing, contemporary cosmology does not seem to worry about the status of its axioms; it is enough that "they work."

However, our research will not end on such a trivial conclusion. In the next part, based on recent theoretical developments, we will explain the precise import of these statements and *prove* why this conclusion is the unavoidable consequence of the scientific method.

NOTES

1. H. Andrillat, B. Hauck, J. Heidmann, A. Maeder & J. Merleau-Ponty, *La cosmologie moderne,* Paris 1984, p. 16. Translation is ours.

2. *Op. cit.* p. 20.

3. Galileo postulated that a body in a state of uniform motion remains in that state unless acted upon by some external agency. Therefore, "Galilean" and "inertial" are synonymous terms in this context.

4. According the postulates of special relativity, all inertial observers (or Galilean frames of reference) are equivalent for the description of movement. In such unaccelerated "inertial" systems (or frames of reference, or observers: all synonymous terms in this context) the principle of inertia holds: all bodies not subjected to any force are either permanently at rest or in uniform rectilinear motion.

5. In two arbitrary Galilean reference systems, temporal and spatial coordinates are related by the Galilean transformation:

$$r' = r - (r_o + vt), \; v = const.$$

$$t' = t$$

where r and r' are the vectors pointing to the moving (point-like) body in the first and in the second frame of reference, v is the vector representing the uniform rectilinear speed of the second frame relative to the first, and r_o is the vector pointing from the origin of the first frame to the origin of the second frame at the instant $t = 0$. The condition $t' = t$ expresses the absolute character of time.

6. "... die Erfahrung [hat] uns die Existenz eines Kraftfeldes (nämlich des Gravitationsfeld) gelehrt [...], welches die merkwürdige Eigenschaft hat, allen Körpern dieselbe Beschleunigung zu erteilen." (A. Einstein, "Die Grundlagen der allgemeinen Relativitätstheorie." *Ann. d. Phys.* 49, 1916; (reprinted in *Das Relativitätsprinzip.* [H.A. Lorentz, A. Einstein, H. Minkowski], Darmstadt 1982, p. 83). Here is our translation: "Experience has taught us the existence of a field of force (e.g., the gravitational field), which has the remarkable property of imparting to all bodies the same acceleration."

7. This fundamental principle can also be formulated as follows: "There is no way, by

experiments confined to an infinitesimal region of space-time, to distinguish one local Lorentz frame from any other local Lorentz frame, in the same or any other region." (Misner, C.W., Thorne, K.S. and Wheeler, J.A., *Gravitation*, San Francisco [Cal.] 1973, p. 386). This important book will be quoted henceforth as Misner *et al.*

8. In a Lorentz frame of reference, free particles are observed to move in a straight line with constant velocity, and the speed of light c is invariant, as is the value of the line element defined as

$$ds^2 = - c^2dt^2 + dx^2 + dy^2 + dz^2$$

9. What follows leans heavily on X. Fustero and E. Verdaguer, "Standard Big Bang Cosmology," in *Foundations of Big Bang Cosmology*, F. Walter Meyerstein ed., Singapore 1989, p. 29-83.

10. As is well known, every scientific model is based on a certain number of presuppositions, and all the elements of the model are logically, i.e., mathematically, deduced from these presuppositions. In modern language, different terms are indistinctly used in a rather careless manner: hypothesis, axiom, axiomatic proposition, *a priori* proposition, assumption, basic postulate, and so forth.

11. Obviously, all the instrumental "extensions" of the human senses must be included. Today, we are capable of "seeing" the universe not only at visible wavelengths, but also at those of X-rays, UV, infra-red, radio, and so on. Furthermore, such indirect "observations" as, for instance, of elementary particle collisions in accelerators, which ultimately reduce to computer print-outs, must equally be included.

12. Plus its matter and energy content. But in some models this matter/energy content vanishes.

13. The rigorous definition of "neighborhood" is beyond the scope of this book.

14. For instance, one can deform, without tearing, a sphere into some elongated body, or a teacup into a donut (or torus), leaving invariant their neighborhood correlations, which thus characterize each (type of) object. This is the subject matter of topology.

15. A topological space X is connected if there are no proper closed subsets A and B of X such that their intersection is the empty set, and their union is in X. In a connected topological space X any two points can be joined by a polygonal line

entirely contained in X.

16. Broadly, the notion of continuity is based on the idea that a function f of the variable x is continuous if, for any small change in x, the corresponding change in $f(x)$ is equally small; or equivalently, in the graph $y = f(x)$ no kinks or breaks appear.

17. If there exists a rule which assigns to each element of a set A an element of a set B, this rule is said to define a mapping function from A into B, written as $f: A \rightarrow B$.

18. We recall that in a Lorentz frame of reference, which is similar to a freely falling elevator, nothing has "weight." In such a local inertial frame, test particles or "free" observers, if not subject to any force, are either at rest or move in a straight line with uniform velocity. "In analyzing physics in a local frame of reference, one wins simplicity by foregoing every reference to what is far away. ... Physics is simple only when viewed locally: that is Einstein's great lesson. ... What is direct and simple and meaningful ... is the geometry of every local [Lorentz] reference frame." (Misner *et al.*, p. 19)

19. The cosmological Big Bang model is assembled so as to represent the entire universe. However, closer to us, at the more modest scale of the solar system, space-time looks Euclidean, i.e., flat, and uncurved, and Newton's theory of gravity has been verified with precision. Accordingly, the mathematical structure of Einstein's general relativity, and of the Big Bang model, must allow the reduction to special relativity in the case of vanishing curvature, i.e., of vanishing gravitational fields, and the reduction to Newton's gravitation in the limit of low gravitational fields, such as are found in our corner of the universe. General relativity and the Big Bang model must, on the other hand, be able to resolve the physics in strong gravitational fields: at the center of galaxies, near a neutron star or a black hole, etc.

20. "Nous considérons qu'aucun fait ne saurait se trouver vrai, ou existant, aucune énonciation véritable, sans qu'il y ait une raison suffciante, pourquoi il en est ainsi et non pas autrement. Quoique ces raisons le plus souvent ne puissent nous être connues." Leibniz, *Monadologie* [1714] 32 (Our English translation).

21. The interested reader will find in M. Heidegger's *Der Satz vom Grund* (Stuttgart 1957) an analysis of the problem posed by this principle along the lines here suggested.

22. Misner *et al.* p. 5.

23. Matter warps space-*time*, and is, in turn, influenced by its geometry. So, space-*time* is no longer Euclidean but curved. We italicize the word *time* to focus on the mind-boggling fact that matter influences time itself! How fast clocks run depends on the matter/energy content of space-time, the extent to which space-time is curved by matter in that place where the clock is ticking.

24. All the successful theories of fundamental physics are varieties of local gauge theories. In such a mathematical theory the imposition of particular space-time symmetries requires the existence in nature of certain forces. The above mentioned principle of invariance is equivalent, in general relativity, to a symmetry under acceleration. Since everywhere a local inertial frame of reference exists, accelerated observers are locally "freely falling" and observe the same phenomena, regardless of their acceleration.

25. Misner *et al.*, p. 42-43.

26. The speed of light is postulated to be a universal constant and limit: every observer, everywhere and everywhen, observes the same value of the speed of light. Furthermore, no information-carrying signal is allowed to travel faster than light (axiom B8).

27. This is due to the universal, and finite, value of the speed of light. Only an infinite speed of light can give meaning to the concept of simultaneity. Only on this condition would any region of the universe be in a position to receive information from (and send to) all simultaneous events everywhere.

28. Space-like geodesics are those paths through space-time followed by particles moving faster than light (tachyons), such particles are forbidden in general relativity. Null geodesics are the paths of photons of light. Time-like geodesics represent the paths of freely falling bodies. Finally, time-like curves with bounded acceleration are the paths along which observers (on rockets, or on the planet Earth) are able, in principle, to move. Such may be the path of event P in Figure 2.14.

29. "Everything which becomes must of necessity become owing to some cause." Note that, quite to the contrary of Leibniz, only the causal temporal *connection* counts, and neither existence nor truth are implied.

30. A cause at the space-time event/point (x, y, z, t) cannot lead to an effect at $(x', y', z',$

t') if a light signal from (x, y, z) cannot reach (x', y', z') in a time span less than $|t - t'|$.

31. They do not occur owing to the stringent symmetries of some models, particularly the standard model we are here considering, but are implied by the theory itself.

32. For instance, a massive black hole is postulated to exist at the center of our and other galaxies to explain the enormous energy released there; these effects may be due to stars and gas falling into such a black hole.

33. The size of the universe is also enormous in the fourth dimension, that is, in the dimension "time." It is generally accepted that the "natural" time unit is the Planck-time: 10^{-43} secs. Current models estimate the lifetime of the universe to be of the order of 10^{60} units of Planck-time.

34. In some conceivable but especially tricky universal space-time geometries, the paths of light rays may be so complex that we would be seeing not an enormous number of different galaxies but only repeated images of a small number of the same objects. In such a mirage-like universe we might be seeing our Milky Way many times over.

35. Simplicity enjoys a special status in cosmology. For instance, in Misner et *al.*, page 1208 we read: "Among all the principles that one can name out of the world of science, it is difficult to think of one more compelling than *simplicity*."

36. Without such simplifications, the determination of solutions to the field equations becomes very complex, if not intractable. Differential equations such as Einstein's field equations represent laws of nature, but we can only observe specific *solutions* of such equations.

37. Isotropy of the universe means that, at any event, an observer cannot distinguish one of his space directions from the other by any local physical measurements.

38. The in principle observable universe may contain something like 100 billion galaxies, which are distant from us all the way up to our "horizon" of 10 - 20 billion light-years away. Furthermore, generally the farther away a galaxy is, the dimmer is its electromagnetic signal which reaches our telescopes and, in consequence, the longer will be the telescope time required for its detection. And, of what we could in principle observe, no more than about 1% has actually been surveyed. Cf. J.O. Burns, "Very large structures in the Universe," *Scientific American*, July 1986, pp. 30-39; and T.J. Broadhurst, G.F.R. Ellis, D.C. Koo & A.S. Szalay, "Large-scale distribution of galaxies at the galactic poles," *Nature*

343, 1990, p. 726-728.

39. That is to simplify sufficiently the Einstein field equations in order to derive observationally testable solutions.

40. Homogeneity means, roughly speaking, that the universe is the same everywhere at a given moment of time. "A given moment of time" is, however, an expression that has no meaning in general relativity, except when one considers such restrictive symmetric constructions as the standard Big Bang model.

41. Notice however that, in this FRW model, all that (dynamically) changes in the geometry of the FRW universe is the scale of distance $R(t)$. All distances between spatial event/points on $t = constant$ hypersurfaces expand (or contract) by the same scale factor. This is a consequence of the postulated isotropy and homogeneity of the FRW universe, and is only true in such a kind of universe.

42. It follows from this equation that $R(t)$ depends on ρ, the density of the matter of the universe. In the perfectly symmetric FRW universe, density, time and the universal scale factor are connected through this simple equation.

43. That is, at $t = t_{big\ bang}$, usually set equal 0.

44. The Doppler effect gives a mathematical formulation to the easily verified observational fact that the frequency of some wavelike signal emitted from a source in movement with respect to the observer appears higher (is blueshifted) in the case of approaching sources, whereas it appears decreased (is redshifted) in the case of receding sources.

45. Redshifts are commonly designated by z, and are defined as the fractional increase in wavelength suffered by a photon emitted at a distant source with wavelength λ_1, and received on Earth with wavelength λ_2:

$z \equiv (\lambda_2 - \lambda_1) / \lambda_1$ [equation 7].

The ratio of $R(t)$ at two distinct times of the history of the universe gives the ratio of the respective dimensions of the universe (cube root of volume) at those two times. It also determines the value $1 + z$. A knowledge of the red shift z experienced in time past by radiation received today is equivalent to the knowledge of the ratio between the universal scale factor $R(t)$ at the time corresponding to z and the value of the scale factor today. Thus, the redshift allows estimating the distance from us of a given redshifted source of electromagnetic radiation.

46. At about 10^{-43} sec. after the Big Bang, temperatures may have been of the order of 10^{32} K, corresponding to densities of the order of 10^{93} g/cm^3. By way of comparison, the highest temperatures that can be obtained in present-day laboratories are of the order of 10^8 K and the density of nuclear matter -i.e., the density of the nucleus of some atom- is estimated to be of the order of $2 \cdot 10^{14}$ g/cm^3.

47. An example of such a symmetry-breaking phase transition can be shown in water. After cooling, one gets ice, where the rotational symmetry of liquid water is "spontaneously broken" and new but more limited symmetries, such as appear in crystal ice, replace the previous structure. However, just as in freezing water the orientation of the symmetry axes of growing crystals may not coincide, and topological "defects" may appear, in the same manner these theories predict the appearance of such defects in the cooling universe. The fact that so far nothing of the sort has been detected, is one of the problems of the FRW model.

48. A black-body is a receptacle that absorbs all the radiation incident upon it, that is, roughly speaking, an oven. After the injection of some energy (heat), radiation in its interior reaches equilibrium: the walls of the oven absorb and reflect heat at some constant rate and a steady state ensues. If one drills a small hole (without disturbing the equilibrium) in the wall of the oven, the outcoming radiation carries energy partitioned in a specific way among the frequencies of such radiation: this is the spectrum of radiation of a black-body. The study of precisely this phenomenon led Planck in 1900 to audaciously propose the quantization of this process in order to arrive at a formula which no classical assumption could justify.

49. It must be emphasized that the abundance of light elements in the universe is not observed, only emission lines from diverse objects such as nebulae.

50. Cf. "Primordial Nucleosynthesis. A critical comparison of theory and observation," by J. Yang, M.S. Turner, G. Steigman, D.N. Schramm, and K.A. Olive, in *The Astrophysical Journal* [The American Astronomical Society] 281, 1984, pp. 493-511.

51. Adducing philosophical, aesthetic, common sense, or other *a priori* "reasons."

52. The operation consisting in finding the minimum set of necessary *a priori* axioms required to give an explanatory sense to some collection of measurements will be interpreted as an "algorithmic compression" of this empirical data set. This operation is almost never feasible, unless of course severe *simplifications in the   obser-*

vational data set are introduced. The FRW model construction, with its drastic simplifications, provides a good example.

53. G.F.R. Ellis, S.D. Nel, R. Maartens, W.R. Stoeger, and A.P. Whitman, "Ideal observational cosmology," *Physics Reports* [Review Section of Physics Letters] 124, 1985, n°s 5 & 6, pp. 315-417.

54. This means that we cannot establish a strict connection between redshift and geometry.

55. In addition to a very crucial "no-news" assumption determining the behavior of the universe model outside our past light cone. Remember that this part of the universe corresponds precisely to the totally unreachable "elsewhere."

56. This minimum can be rigorously measured: the complexity of this minimum set of *a priori* propositions can in no case be inferior to the complexity of the observational data set; cf. the third part of this book.

57. In the case of open universes, negatively curved or flat, which are spatially infinite, the information containing volume of our past light cone literally represents zero percent of the total. Things get even worse in the inflationary scenario described below, where "our" universe is only one amongst many other existing universes.

58. "If, then, anyone can claim that he has chosen one [model] that is fairer [better] for the construction of these bodies, he, as a friend rather than a foe, is the victor" (*Timaeus* 54a-b)

59. In the present universe, also called "matter dominated universe," this density can be measured in "atoms per cubic meter," at about 0.1 atoms of "luminous" i.e., visible matter (see below) per cubic meter. By contrast, the early universe, before the matter/radiation decoupling epoch, is designated as the "radiation dominated universe."

60. An enormously large time measured in the time-scale units of particle physics: 10^{60} units of Planck-time.

61. Cf. V. Rubin, "The Rotation of Spiral Galaxies," *Science* 220, 1983, p. 1339-1344.

62. Some dark matter candidates: massive neutrinos, weakly interacting massive particles (WIMPs), axions, photinos, gravitinos, mini-black holes, etc., all undetected so far.

63. Cf. L.F. Abbott and So-Young Pi, ed., *Inflationary Cosmology*, Singapore 1986.

64. S. Hawking wrote, "The fact that we have observed the universe to be isotropic is only a consequence of our existence." C. B. Collins & S. W. Hawking, "Why is the Universe Isotropic?" *The Astrophysical Journal* 180 (1973) 317-334.

65. One MeV = 106 electron-Volt. Masses in particle physics are given in energy units. 1 MeV/c2 = 1.78 x 10-30 kg.

66. All protons have a positive electric charge, identical in value to the electrical - but negative - charge of the electron, and so in principle they should all be subject to a repelling Coulomb force. What makes the nuclei of atoms stable is thus the strong force.

67. The neutron, when bounded in an atomic nucleus, is considered to be as stable as the proton.

68. The possibility of the decay of the proton is postulated by most of the Grand Unified Theories. How can one measure such a long time? What is done is to search for such a decay event among many protons, that is, in very large volumes, hoping to find, statistically, one telltale proton decay event. So far, without result.

69. The Lagrangian function is a set of second-order differential equations for a system of particles which relate the energy of the system to its (generalized) coordinates, the (generalized) forces acting in it, and time. There is only one equation for each of the degrees of freedom possessed by the system.

70. This quantity is given by the generator of the symmetry group.

71. Symmetry is conceptually quite close to the notion of pattern. In a pattern, something is repeated, the same element can be discerned over and over, as if on a wall paper design. Cf. Hermann Weyl, *Symmetry*, Princeton 1952. In this book, Weyl defines symmetry as "a harmony of proportions," and he reminds us of the very revealing German word *Ausgewogenheit* (approximate meaning: well-balanced). In another important remark, Weyl states that the theory of relativity is nothing but an aspect of symmetry! In the final part of our book, still another crucial aspect of symmetry will be brought to light: it represents the exact opposite of the notion of *complexity*. And since complex phenomena escape rational analysis, symmetry remains the primary condition for any acquisition of *knowledge* about the physical world. This absolutely central fact was clearly pointed out by

Plato, twenty-four centuries ago.

72. Lie groups are related to the concept of "differentiable manifold" mentioned above.

73. Plato already understood the notion of symmetry in terms of some primordial mathematical group theory, although this theory was formally introduced only in the early nineteenth century. Nevertheless, since he wanted to give his elementary particles maximal symmetry, Plato made them correspond to the regular polyhedra, i.e., solid shapes whose sides are identical polygons. These polyhedra, as well as the sphere, the shape chosen for the entire universe ("spherical symmetry," as in FRW), have the property of remaining invariant under rotations: the sphere under any rotation, the cube under 90° rotations, the tetrahedron under 60° rotations, and so on.

74 . I.e., something measurable in the real world.

75. In quantum mechanics, eigenvalues of the state function describing a physical system are the only values of a linear superposition of possibilities that can eventually be observed, with probabilities given precisely by that state function.

76. Proton rest mass $= (1.67252 \pm 0.00008) \cdot 10^{-27}$ kg.
Nucleon rest mass $= (1.67482 \pm 0.00008) \cdot 10^{-27}$ kg.

77. The masses of the particles in a multiplet have quite disparate values.

Part III

What Knowledge is Conveyed by Science?

We, living today, have become entangled, by the peculiar dominance of contemporary science, in the bizarre misconception that science constitutes a source of knowledge, and that thinking must yield to the judgment of science. But whatever a thinker is capable of expressing can never be logically or empirically confirmed or refuted. Nor is it a matter of belief. It can only be faced in a questioning-thinking manner. And whatever is faced in that way is invariably that which is *worth* questioning.

This particularly dense and poignant text by M. Heidegger,[1] here freely translated, affirms that "true" knowledge (*episteme*) can never issue from the natural sciences as they are nowadays understood, and that the only remaining available option is the unrelenting reformulation of the questions, the ceaseless restating of the questions *worth* asking.

This being so, to what does the knowledge conveyed by the science of nature ultimately amount? Where are its limits? The ancient "problem of knowledge," here formulated by Heidegger in the terminology of traditional philosophy, constitutes the subject we will now address. But our strategy departs from traditional methodology in that we approach this question in mathematical terms derived from recent *Algorithmic Information Theory*. By this means, the questions *worth* being asked, and *what* is thereby questioned, will now become clear and precise.

The Symbolic Description of Reality

The problem of *defining* what kind of knowledge is conveyed by the scientific approach, the problem of clarifying what is *explained* by this method and the problem of *interpreting* the scientific results: all these problems were already raised in Plato's *Timaeus*. In the solutions to this problem of knowledge as set forth by philosophers, a number of indispensable assumptions are in general implicitly included.

1)   The theoretical organization of the physical world follows a very specific method, an epistemology, developed for the first time in history by Plato and described in the *Timaeus*. This method is founded on this primordial conviction: *only a mathematical or equivalently a logical demonstration can lead unerringly to a scientific knowledge.*[2] This conviction entails:

2)   *A scientific inquiry is based on a system of assumptions, provisionally admitted as the foundation of a theoretical model of the physical world: the axioms.* The axiomatic systems underlying the cosmology of Plato's *Timaeus* and the standard Big Bang model sufficiently illustrate this important point.

3)   It follows that an analysis of the knowledge disclosed by the scientific method must focus on the axiom system admitted in each case. But *asking questions about these axioms is equivalent to appending additional axioms* to the system. Therefore, the initial question can be reformulated. Instead of asking: *What knowledge does science convey?* we now ask: *What knowledge do the axioms convey?*

4)   Finally, to reach scientific knowledge, to put forward the required list of axiomatic propositions, it is necessary *previously* to have replaced concrete sense-data with their corresponding *symbolic representation*, according to theoretical rules accepted

as true and valid by scientists. A complex procedure allows experimentally obtained sense-data to be replaced by *measures*.

5) The scientist's aim, when studying a particular physical system, is to uncover the specific correlations that may prevail amongst these abstract notions, the measures. For this manipulation of abstract symbols, mathematics are essential and irreplaceable, which explains why modern physical science has become predominantly mathematical. Clearly, the final objective of such research consists in finding a mathematical theory, a formal axiomatic system, in which proved theorems can be equated with predictions concerning possible measurements to be performed within the system under consideration.[3]

The *Timaeus* and the Big Bang model provide two conspicuous examples of this method. They make clear how such a formal axiomatic system is assembled, how the universe is mathematically *invented*. Underlying this method, however, are a certain number of crucial *prejudices*. Some have been already mentioned in the second part; here we focus on those relevant to the ensuing stage of our analysis:

1) The world must present a sufficient degree of order so that the measurement operation can be performed.[4] This prejudged *stability and invariance*[5] implies:
    - the possibility of distinguishing objects;
    - the possibility of naming them; and
    - the possibility of measuring.

2) Names and measures are abstractions. They are not given directly by sense-data, in an unmediated fashion.

3) The only effective method of analysis consists in translating such abstractions as names and measures into pure symbolic forms in order to operate with them *mathematically*.

4)   As a result of the symbolic representation of sense-data, a mathematical theory can be established, allowing theorems to be mathematically *deduced* from presuppositions assumed to be true. Such theorems relate to physical measurements. Hence, the theory represents a model of reality. Thanks to this "artifice," a mediation between sense-data and intelligible intuition can be established. It is further assumed that this deductive procedure must follow a set of *rules of inference* presumed to incorporate the basic prejudices of human logic.[6]

5)   In fact, it is admitted that the logic underlying the rules of inference is such that this deduction can be achieved by calculus, *mechanically*,   unambiguously,   and   furthermore,   leads mechanically to *certainty*. In modern terminology, such rules of inference mechanically connecting (*mapping*) axioms and theorems are called an *algorithm*.

6)   Once these prejudices are accepted, a definition of a *scientific method*, based on a formal axiomatic system, allowing us to *invent the universe*, can be reached. Succinctly, the essential components of this *formal axiomatic system* are:
   • the explanation of its symbols (the dictionary);
   • the rules governing the arrangement of these symbols into valid propositions (the syntax);
   • the set of valid propositions *a priori* admitted as true (the axioms);
   • the rules of inference (the algorithm) by which valid propositions of the system can be arranged into additional valid (true) propositions (the theorems);
   • a decision procedure establishing which of the valid propositions is true and which is false (i.e., a mechanical procedure determining whether a proof of a theorem follows from the admitted rules or not).[7]

Two theoretical systems, as far apart in time as the *Timaeus* and the Big Bang, will now be studied: Aristotle's *Posterior Analytics*, and *Algorithmic Information Theory*. For the last, we rely entirely on the work of Gregory J. Chaitin.[8]

Algorithmic Information Theory: Some Relevant Aspects[9]

In the preceding discussion on cosmological models, we have time and again stressed that what really matters in any such construction are not the particular coincidences linking the theoretical predictions (i.e., the theorems) of such models with observationally collected data but, on the contrary, the set of their axioms, the list of their *a priori* assumed propositions.

This claim can be rigorously *proved* when reformulated in terms of the Algorithmic Information Theory. In fact, this approach leads directly to Gödel's famous "incompleteness theorem," providing evidence that the incompleteness phenomenon discovered by Gödel is natural and widespread rather than pathological and unusual, that it applies each time the deductive method is employed, and thus can not so easily be swept under the rug.[10]

Gödel showed that, in any axiomatic formal system capable of deducing the theorems of arithmetic, there are theorems that are true but that can never be proved or disproved in this system; that is, they are undecidable.[11]

Algorithmic Information Theory represents a spectacular extension of Gödel's theorem. In the "algorithmic" version, Gödel's theorem becomes:

If a theorem of some formal axiomatic system contains more information than the set of axioms of such a system, then it is impossible for the theorem to be derived from these axioms; such a theorem is then undecidable (in this system).

Paraphrasing Aristotle, we may say that all the information, all the theorems that can be mechanically (algorithmically) inferred within a theoretical model, appear to be entirely contained as a potentiality in the axiom system of this model; the development of the theory starting from the axiom system does not add any new information! On the other hand, all the uncountable infinite propositions containing more information than a particular axiom system can never be proved by that system; these *questions* simply cannot be answered by that system, and remain undecidable. By the same token, any extrapolation going beyond the information content of the axioms is either undecidable or must be added as a new axiom, without proof. In other words, any *question* about the axioms can only be answered in a metasystem, a system enlarged by at least that question. Questions *about axioms* are in this context "unscientific," since they do not belong to the scientific theory being considered. Moreover, a mechanical decision procedure, establishing which of the valid propositions of a formal system are true and which are false, cannot exist.

In summary, formal axiomatic systems such as cosmological models and physical theories are limited; they are *incomplete*, in that an indefinite number of pertinent valid questions remains undecidable. How to measure the power of a formal axiomatic system or, more to the point, how to measure its degree of incompleteness, is the next question to be elucidated.

A formal axiomatic system is judged to be equivalent to the sum of all its possible theorems, deduced from its axioms following its admitted rules of inference. And the basic contention of Algorithmic Information Theory is that one can study these systems considering the axiom system as the *program* of some kind of computer, to be defined, such that if one runs the axiom system as an input program, this computer will produce as its output, using the rules of inference as its algorithm, the whole theory, that is, all its potentially provable theorems.

Turning the question around, we assume as given some set of provable theorems of a theory.[12] In terms of Algorithmic Information Theory, this set of theorems can be considered to be the output of some computer to which the respective formal axiomatic system (axioms plus rules) has been given as input. It now becomes possible to focus on one individual object: that particular output, the set of theorems, which, if coded in binary digits (*bits*),[13] will be some string of 0's and 1's. The question then becomes: given one particular string of 0's and 1's, which we now assimilate to theorems of some theory in the above sense, what is the size in bits of the *shortest* program for generating it as output in the described manner? What is the shortest program for calculating this string? Or, what is the most compact formal axiomatic system from which these theorems can be deduced? If an answer to these questions can be found, and assuming the rules of inference as known, one can evaluate the minimum of *a priori* information that must be accepted without proof for some theoretical construct, a result of obvious importance.

To proceed, a certain number of *definitions* are necessary.

The *algorithmic information content I(X)* of an individual object X (some string of 0's and 1's) is defined to be the size of the smallest program outputting (calculating) X.

The *joint information I(X,Y)* of two objects X and Y is defined to be the size of the smallest program able to calculate X and Y simultaneously.

And the *mutual information content I(X:Y)* measures the commonality of the two objects: it is defined as the extent to which knowing X helps one to calculate Y.

The next two notions will play a fundamental role in our discussion. They are *algorithmic randomness* and *algorithmic independence*. These concepts are quite close to the intuitive notions of disorder, lack of any discernible pattern, chaos, and totally unrelated objects, respectively.

We now consider some string N-bits long. Most of such strings have an algorithmic information content $I(X)$ approximately equal to N plus the algorithmic information contained in the number N itself which is of the order of log_2N. That is, to output such a string on a

computer, one generally needs an input program essentially equivalent
to the bit-length of the string itself. No string N bits long can have a
greater algorithmic information content, i.e., greater than $N +$ *(order
of)*$log_2 N$. A few strings, however, may have a substantially smaller
information content: they are the strings which have some regular,
repetitive pattern, some symmetry. N-bits long strings with information
content of about N are said to be *algorithmically random*. This
definition applies identically to infinitely long strings. For instance, the
number π is the extreme opposite of an algorithmically random string:
it takes only a small program to calculate the infinite digits of π,
although the task takes infinite time. This factor is of no relevance
here, due to the type of computer used, as will be described below.
But, if one asks what is the *nth* digit in the decimal expansion of π, a
program much shorter than n can give an exact result in finite time.[14]

It follows that an algorithmically random N-bits long string is a
string lacking any kind of pattern, such that to create such a string in a
computer requires an input program approximately similar in size to
that of the string itself. This means that if one has such an N-bits long
string and wishes to predict the $N+1$ bit, one cannot predict it with a
probability better than $\frac{1}{2}$, the probability of guessing the outcome of
the toss of a fair coin. Alternatively stated, any additional bit one adds
to the string requires exactly one bit of program. In contrast, if the
string can be *algorithmically compressed* into a (substantially) shorter
program, that is, if its algorithmic information content $I(X)$ is much
smaller than N, which will be the case of a string which shows some
pattern, as for instance

00111001110011100111.....

which can be algorithmically compressed into a short program of the
type
"repeat 00111 x-times,"

then the prediction of the $N+1$ bit can be made with much greater
accuracy (certainty, in this case). On the other hand, it is evident that

the probability of an infinite sequence obtained by independent tosses of a fair coin to be algorithmically random is equal 1.

To recapitulate, algorithmically random strings are sequences that cannot be algorithmically compressed into input programs substantially shorter than the sequence itself. Algorithmically random strings contain incompressible, irreducible information. It follows that the concept of algorithmic information closely corresponds to the concept of complexity: *the algorithmic information content I(X) of a sequence of binary symbols (bits) measures the complexity of this sequence.* Algorithmic information content and (algorithmic) complexity of some binary string are synonymous terms in this context. What specifically makes Algorithmic Information Theory pertinent to our investigation is the *equivalence between a formal axiomatic system and an algorithm.* More precisely, since the particular computer here used can be mathematically defined, it becomes rigorously possible to consider the axioms and the rules of inference of a formal axiomatic system to be encoded as a single finite binary string, constituting the input of this computer. And *the algorithmic complexity of a formal axiomatic system determines the limit of the complexity of the theorems that can be deduced.*

The Turing Machine

We now must define the "computer" on which the above described operations, in particular the algorithmic compression, are implemented. This *Gedankenexperiment* is called a "universal Turing machine,"[15] created by the British mathematician Alan Turing in the 1930s.

The Turing machine is based on the idea that any *method* is an algorithm, a set of instructions expressed in some formally defined language, which enable a machine to mechanically carry out some program. This imaginary machine can be considered to be, at the most fundamental level, the embodiment of mathematical reasoning, but it also marks off the limits of computability. Actually, given a large but finite amount of time, a Turing machine is capable in principle of performing any computation that can be done on any modern computer, no matter how powerful. Moreover, combining "simple" Turing machines into a "universal" Turing machine, this imaginary machine can in principle carry out any task that can be explicitly described. Such a universal Turing machine can simulate any operation performed on any particular Turing machine. Consequently, what a Turing machine cannot compute may represent an ultimate limit of the power of mathematics.

A simple Turing machine is schematically shown in the Figure 3.17. In a way, it reminds us of a typical typewriter. It incorporates a moving printing head that prints discrete symbols, for instance 0's or 1's, one at a time, on a infinitely long printing tape, marked off into discrete frames or segments. Only one symbol is allowed in each frame. The printing head can carry out two other functions: it can remove or erase one symbol at a time from the tape, and it can "read," or register the symbolic content of each frame, one frame at a time. In this way the symbols on the tape can be considered to be the input to the machine.

Furthermore, the machine can assume any one of a finite number of states, according to the configuration of the machine, here considered as a sort of "black box" for which no mechanical description is given. It is only required that each state be described in terms of the effect that state has on the activity of the machine. Thus, the Turing machine functions in discrete, instantaneous steps, and each sequential step is determined by the current state of the machine plus the symbol (in our example: a 0 or a 1) that occupies the frame currently scanned. These two initial conditions determine at each discrete step a set of three instructions:

First, the machine is instructed what to do with the symbol on the frame being currently scanned: it may leave it unchanged, or replace it and print another symbol.

Second, depending on the scanned symbol and the current state of the machine, the next state is designated: the machine may change to another of its limited set of states or remain unchanged.

Figure 3.17

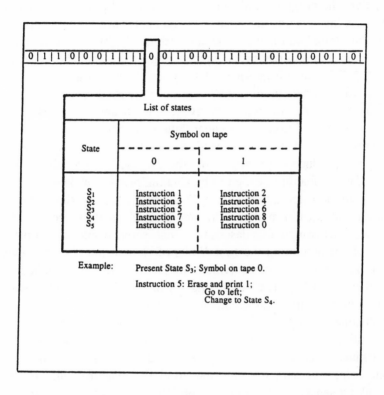

Third, the printing head is instructed to move either to the frame on the left or on the right, or not to move at all.

These instructions, contained in a finite list of states, are all that is needed to perform any possible kind of computation, since any particular Turing machine, having just a limited (finite) list of states and a limited alphabet of symbols, can itself be encoded on a tape as an input sequence of 0's and 1's of a universal Turing machine; this universal Turing machine will then perform in a way identical to that particular machine.

Algorithmic Information Theory presumes the existence of such a "computer:" the symbols that are printed at some point in time on the tape are the program, and running this tape will produce some output, if – and this is a crucial aspect – the computer ever halts. Suppose that the "program" on the tape is created arbitrarily, by tossing each time a fair coin and so creating an input string of 0's and 1's. What is the probability that, given such arbitrary input, the Turing machine will halt after some finite time? Turing proved this question to be *undecidable*: it is impossible to give a concrete answer to this question. In Algorithmic Information Theory, the undecidable question: What is the probability that a computer given an arbitrary input will eventually stop and produce some output? has the answer: This probability is Ω.

This probability, called Ω by Chaitin, plays a fundamental role in Algorithmic Information Theory.[16] Because it can be shown that Ω is an algorithmically random ("incompressible") real number which has no compact computable description, that is, one cannot find a program capable of outputting the first N bits of Ω in a time T_N, whatever N. Put differently, Ω is equivalent to an infinite set of independent mathematical facts (i.e., each single 0 or 1 of its binary expression is an independent mathematical element), and this irreducible mathematical information cannot be compressed into any finite algorithm. Since Ω measures the probability that a universal Turing machine, given an input string generated by independent tosses of a fair coin, will halt, the question: Will the machine halt? is strictly speaking senseless. The only option left is to run that program on the machine and wait... This is "algorithmic undecidability".

According to theorems proved in Algorithmic Information Theory, the following assertions, directly pertinent to our analysis, can be put forward:

• Given some sequence, such as an N-bits long string of 0's and 1's, one can never prove the randomness of such a string. Such a string may be created by arbitrary tosses of a fair coin, but it may also correspond to the symbolic description of some physical situation.

Whatever the origin of the string, the question: is such a string algorithmically random? is undecidable. On the other hand, one can show that such a sequence is non-random only by discovering a short algorithm to generate it. Lacking this short algorithm, either the sequence is totally devoid of any pattern, and therefore algorithmically incompressible, or one must keep trying, hoping eventually to find one. But no *proof* of randomness is possible.

• If one picks any arbitrary infinite sequence, for instance, the binary infinite expansion of some arbitrary real number in the interval between zero and one, this number will be, with probability equal to unity, that is with certainty, algorithmically random.

• If one asks, what is the probability that some finite binary string can algorithmically be compressed by just 10 bits, the answer is that about 999 strings out of 1000 cannot be compressed even by such a meager amount! This means that the immense majority of the possible sequences are not algorithmically compressible, and cannot be generated by an algorithm containing less information than the sequence itself.[17]

• From the last observation, one might deduce that it is the easiest thing in the world to exhibit a specimen of a long series of random digits, in order to print a truly algorithmically random number. However, the contrary is the case. It is impossible to do so, a consequence of the theorem which states that one can never prove some sequence to be algorithmically random, that such a question is undecidable.

• To come back to the main line of our argumentation: suppose we now consider a formal axiomatic system to be a computer program for listing its. *complete* set of theorems, and suppose we wish to measure its size in bits. A crude measure of the size in bits of a formal axiomatic system is the amount of computer space it takes to specify a proof-checking algorithm and how to apply it to all possible proofs, which is approximately the amount of space it takes to be very precise about the dictionary, the syntax, the axioms and the rules of inference. This can be considered to be roughly proportional to the number of pages it takes to present the formal axiomatic system in a textbook.[18]

• Based on this complexity measure of a formal axiomatic system, the following incompleteness theorem can be proved. Consider an *N-bit formal axiom system*. Then there is a computer program of size *N* which does not halt, whose output is undecidable. But one cannot prove this undecidability within the axiomatic system. Moreover, within an *N*-bit axiomatic system, it is not even possible to prove that a particular object has algorithmic information content greater than *N*, even though almost all objects have this property. Put more generally: the information content (the complexity) of the output of some program of complexity *N*-bits, corresponding to the theorems deducible from an axiom system, cannot be greater than *N*-bits, which means that they cannot contain more information than the axiomatic system itself. This fundamental conclusion is just Gödel's theorem, and Turing's halting theorem, in its more general algorithmic form.

Armed with these insights provided by Algorithmic Information Theory we now return to our subject and analyze their application to our discussion.

A definition of science

We start by asking the question: what is science?

From our specific viewpoint, a scientist starts, by means of the definition of abstract concepts of measuring standards and the consequent possibility of application of the measurement operation, by reducing his observations of some local physical situation or arrangement to a quantitative description. This description is then translated into a purely symbolic formulation such as a sequence of binary digits, 0's and 1's. That such a reduction of sense-experienced reality to some sequence of symbols is always possible is the essential presupposition of our analysis. Moreover, such sequence may presumably be very large.[19]

The scientist then examines increasingly larger segments[20] of this enormous sequence, with the aim of finding the shortest possible computer program which may compute some, possibly infinite, binary sequence whose initial segment reproduces to acceptable accuracy the "observed" segment. This computer program, this algorithm, is the scientific theory, that is, the corresponding formal axiomatic system, and the output of such a program is a verisimilar account in the Platonic sense. However, to the extent that the scientist has been able to algorithmically compress the "observed" sequence into his "program", into his theory, predictions (theorems) can be advanced, and experimental tests can be performed. They will show the coincidence to some acceptable degree of accuracy between segments of the two strings: the string "theorems", that is, the predicted quantitative measures, and the string "reality", that is, the actually observed measures. It then becomes feasible to *choose* among competing theories on the base of the best fit between theorems and observation. These ideas are illustrated in Figure 3.18.

Figure 3.18

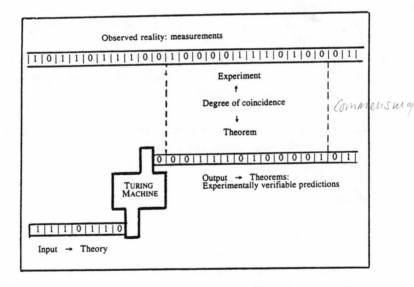

Note: In many descriptions of the Turing machine, only a single tape is shown. Here, to make things clear, the tape is split into two separate tapes. Initially, only the theory is printed on the lower tape. After some time-lapse[21] and after the machine has finished scanning the entire theory-tape, it will print the "result" on the theorems-tape. The observed reality-tape represents real measurements which, as assumed, are always expressible in binary code. The degree of coincidence between a segment of the theorem-tape and the chosen segment of the measurement-tape is what confirms or refutes the theory.

Scientific theories will thus be selected following at least three criteria:

1) shortness of program, that is, extent of algorithmic compression of some particular sequence, of some chosen segment of the "observations" string, the shortest algorithm being preferred;

2) degree or accuracy of coincidence between the predicted and the empirically observed values;

3) length of the sequence for which some degree of coincidence can be established, i.e., the scope of the exterior reality encompassed or "explained" by that theory.

In keeping with this "algorithmic" definition of science, some additional remarks and some preliminary conclusions can be put forward.

1) Measurability

In science, and more precisely in physics, a theory culminates in the formulation of some system of differential equations, such as Einstein's field equations described in the second part. The problem then becomes to implement the theory, that is, to solve these equations. This requires the previous determination of initial and boundary conditions which in turn can only be obtained from experiment, from a measurement. In other words, the measurable numbers of the theory must be computable in the algorithmic sense. Clearly, it must be feasible to generate them mechanically by means of some machine, be it a computer coupled to a measuring device outputting a *number*, or a Turing machine. But we have seen that most numbers are algorithmically random. What happens now? We quote without further comment R. Geroch and J.R. Hartle:[22]

> Imagine a physical theory whose prediction, for the result of some particular experiment, is the non computable number K. ... There exists no algorithm to determine whether a given number is within [some arbitrary value] ε of K – for this is the essence of K's being non computable. Of course, it is nonetheless possible to test the theory – and to do so to arbitrary accuracy. The point is only that the test cannot be carried out mechanically: each new level of accuracy will require new ideas for the testing procedure. On the other hand [they continue], imagine that one had access to experiments in the physical world, but lacked any physical theory whatsoever. Then no number could be shown to be measurable. ... It is only a *theory* that can guarantee [the measurability of some number].

2) Universal Turing Machines

To show that some binary string presents an algorithmic complexity of m bits, it is necessary to have at one's disposal axioms with complexity of at least $m + c$ bits, where c is a constant independent of the axioms; nothing less will do. Therefore, within the framework of some specific formal axiomatic system, it is not possible to prove that some string is of complexity greater than some fixed value. *A fortiori*, it is impossible to prove if an infinite – or very long – string is compressible or not. All these ascertainments are based on the use of universal Turing machines in the proof of the theorems. It is however not excluded that, for a *specific* very long but finite string, an

Inventing the Universe

equally specific Turing machine can be constructed with its finite list of states so designed, that by simply pressing a button, or writing a *1* as input, this machine outputs the entire string. Based on the definition of algorithmic information content or algorithmic complexity, it seems that this string has minimum complexity: the program outputting that string on a Turing machine is only one bit long – even if the string is very long and algorithmically random, i.e., of maximal complexity! And since this operation can be repeated for every string, where does this manifest contradiction leave our definitions? What happens, in fact, is that the complexity of the string has now been incorporated into the "hardware" of the specific machine: the more complex the string, the more states must be included in its list. In the worst case, the entire string is included, plus the instruction: "copy this string when the input is *1*." The specific machine ceases to act as a computer and reduces to a Xerox machine. By introducing *universal* machines we recover our initial definitions. In effect, the universal machine must equally be able to output the string in question. One must write on the input string of the universal machine the specific program of the specific machine, that is, only *1*, but furthermore, one must write the complete description of that particular machine. And that description will be, if our initial string is of some complexity x, at least as long as x bits. In consequence, the minimum program of the universal machine for outputting the initial string will correctly reflect by its length, which now includes the description of the specific machine, the complexity of this string.

3) The Brain and Machine-Language

Could our brain resemble such a *specific* Turing machine, designed (by evolution) to interpret as simple some enormously difficult tasks, whereas others are considered to be of unlimited complexity? It can be shown that in our brains the portion allocated to the recognition of human faces is comparatively very large, representing our capacity of instantly differentiating such subtle, but vitally important, expressions as love, hostility, respect, etc. Face-recognition seems to be incorporated in the brain's "hardware," so that it appears (to us) easy and simple to accomplish what is in essence an incredibly difficult task. Such considerations lead to the question: which is the particular Turing machine we ought to select in order to apply Algorithmic Information Theory to the special case of the "problem of knowledge"? In other words, what is under these circumstances the appropriate machine-language? Chaitin suggests a modified version of LISP. Actually, the only condition we demand from the Turing machine is to represent a mechanical procedure to infer, in some formal system, the theorems from the axioms, that its list of states be capable of encoding, in some way we have not established, what corresponds to the currently known mathematics and logic. It remains an open possibility that, if in the future a different mathematics or logic can be found, a different universe might be invented.[23]

4) What can we know?

Our approach takes for granted that – at least in principle – it should always be possible to encode the sense-data observationally gleaned from the physical world on some, presumably gigantic string of 0's and 1's; and that what is usually called "the explanation" of those observations is conceivable if and only if this string can be compressed into a substantially shorter algorithm. But Algorithmic Information Theory proves that it is impossible to ascertain *a priori* if the requested compression is feasible. In fact, the irreducibly undecidable character of this probability is reflected in that horrendous number Ω. Consequently, when analyzing a set of observations from nature, that is, when studying some selected parcel of the universe, it is impossible to establish if eventually an explanatory theory will be found *before* a theory having the required predictive power has effectively been validated. In other words: *there is no method for mechanically answering this question: what can be known?* The only remaining alternative, as pointed out by Heidegger, consists in reposing the questions, suggesting some new axioms, and then finding out if the theorems deducible from them are adequate, that is, if "it works."

Nothing then allows us to contend that the universe, or a part of it, is ordered and symmetric *before* the algorithmic compression of the string describing it has been achieved. But the probability of achieving a compression of only 10 bits is just one in a thousand. More precisely, the probability that a string n-bits long can be algorithmically compressed into a $n - k$ bits algorithm has shown to be of the (order of) $(\frac{1}{2})^k$; hence this probability decreases exponentially as k increases. Since the number of existing possible programs of $n - k$ bits length decreases exponentially with k, a shorter string cannot be assigned to the immense majority of the 2^n possible strings representing the descriptions of some physical situation: in other words, these strings cannot be compressed, and so can never be "explained." Consequently, the descriptions encoded as incompressible strings are algorithmically random, devoid of any order or symmetry.

Complex problems, for instance those found in biology, inescapably require gigantic strings for their *complete* description. Evolution has shaped the biosphere by acting for four billion years on hundreds of chemical reactions taking place every minute in almost every cubic centimeter on the surface of the Earth. To arrive at an "explanatory theory" taking account of this process, the "mega-string" representing it must be compressed to some size manageable by humans; say, the size in bits of ten thousand text books. Immediate corollary: this is so highly improbable to be considered strictly impossible. In other words, when contemplating the complexity of the phenomena occurring in our "sublunary" world, the string necessary for its complete description is so gigantic, and our human capacity so limited, that *an explanatory theory globally accounting for our complex world must be considered absolutely unattainable.*

Experience has shown that locally, specific, isolated phenomena can be approximated by mathematical theories, sometimes even with astonishing precision. But a global approach, as required for instance by cosmology, can never escape these conclusions unless, of course, the simplicity, the order, and the symmetry are *a priori* postulated to be the result of some pre-established fundamental mathematical harmony pervading the world (the result of the intervention of a considerate demiurge?): only then might the compression of this string to human-manageable sizes become feasible. Indeed, this is exactly how the Big Bang model was formulated in the first place. Arguments purporting to support the conviction of a *universal* order such as "the sun rises daily" are worthless in this context, which derives as much from Hume's remarks as from the neglect of the obvious scale differences between a human day and astronomic time.

In the *Timaeus*, it is presupposed that a benevolent demiurge puts up his best effort to introduce order into the world (*Timaeus* 29c-30d). In the same vein, it is conceivable that some all-powerful divinity might have conveniently arranged the bits of this "sublunary mega-string" precisely suiting the capabilities of the present human mind. Or less contrived, as Aristotle believed, the universe may be teleologically ruled, and everything in nature happens with some purpose in view (*Physics* II 8, 198 b 16), – this purpose being precisely the compressed algorithm we have been looking for: in this universe nature never acts in vain (*De caelo* I 4, 271 a 33). So many things are "explained" if this is accepted, the world becomes so nice and clear, that, not surprisingly, most philosophers have followed Aristotle: Descartes, Newton, Leibniz, Kant, etc. In more recent times, a contested and somewhat dubious "Anthropic Principle" has been evoked, asserting that the universal mega-string is not algorithmically random simply because "we are here to observe the universe." Finally, it might be assumed that an universal evolutionary process, allowing at the present epoch the appearance of conscious "ordered" observers, must necessarily entail a sufficient degree of "harmony and proportion" at least in our corner of the universe, and at time-scales compatible with human observational possibilities. But this assumption only displaces the problem: if a compression by only 10 bits is already so highly improbable, and if all teleological or similar arguments are discarded as unscientific, one must now explain this astonishingly selective universal evolutionary process, this "fine tuning" to precisely our present-day proficiency.

5) What do we know?

But modern science has repeatedly and astoundingly shown that in many situations an algorithmic compression *is* feasible, that a formal axiomatic system, a theory, is capable of outputting a significant string of data whose predicted numerical values are in adequate agreement with observation and experimentation. In some cases, this agreement is impressive. A frequently adduced illustration is the value of the magnetic field of the electron, the experimental result being greater by a factor of exactly $1.001159652188 \pm 0.000000000004$ than the value predicted by quantum electrodynamics![24] And despite its disturbing remaining problems, the Big Bang model can also claim to have observationally verified certain key predictions, as we have pointed out. In all such instances, in what exactly does the knowledge thus acquired consist?

A theory is equivalent to the shortest algorithm into which the symbolic string describing some particular parcel of the universe has been compressed, provided that it agrees reasonably well with predicted observations, provided that "it works." Should that theory be amenable in its turn to further algorithmic compression, resulting in a still shorter algorithm, then this further compressed theory will be considered to be "the" theory, since it can output the initial longer theory just as well. If the compression results in the minimal algorithm, then the information content of that maximally compressed string contains all the information of the theory, that is, contains potentially all its theorems and consequently contains all the predictions amenable to verification that this theory can ever establish. Theorems containing more information than the minimal algorithm can never be proved in the framework of the theory: they are undecidable and correspond to questions about the axioms, and are therefore equivalent to additional new independent axioms.

Let us now assume that an algorithmic compression of the symbolic description of a segment of nature has been successfully accomplished and that a theory has been established: in what does the knowledge gained by this procedure consist?

The extent of such knowledge reduces to a (compressed) algorithm, to a formal axiomatic system. Until further compression is achieved, one considers the binary string representing this system as the shortest possible one, or the minimal algorithm. Regardless of the fact that such an assumption can never be proved, if in its present formulation the algorithm cannot be further compressed, if it is considered to be the shortest possible one, then inescapably it follows from Algorithmic Information Theory that this string, this algorithm, this theory, is – by definition! – algorithmically random, incompressible, strictly irreducible and devoid of any order or symmetry. No logic, no Turing machine, can *deduce* our theory from a more concise, more compact theory, or from more fundamental principles. The only possible manipulation consists in copying the theory, like in a Xerox machine, since the minimal program for outputting the theory has approximately its same length in bits. No logical analysis of such a theory is ever possible, and the only attainable operation reduces to its simple enunciation.

This surprising conclusion is the direct corollary stemming from our definitions. The shortest algorithm, the most compact program is necessarily algorithmically random. Otherwise, this *shortest* program would admit a compression into an even shorter one: a manifest contradiction. What does "algorithmically random" mean? It means that the theory, expressed as a binary string of 0's and 1's, is such that if one arbitrarily picks one bit, the chance of hitting a 0 or a 1 is ½, and if you want to bet on whether the next bit is a 0 or a 1, there is no procedure more effective than tossing a fair coin. Since, considered in the light of Algorithmic Information Theory, formal axiomatic systems are algorithmically random, they are irreducible to any compact description, and cannot become the object of a rational analysis; they cannot be reduced to more primordial or simple elements, but can only be enunciated.

Anything surpassing the algorithmic information content of the system, any theorem surpassing its complexity, cannot be deduced by the theory. Incorporating *new* information in a theory is equivalent to adding axioms. This is what usually happens when aberrant observations, counter-evidences or anomalies not predicted by the theory are incorporated *in* the theory: this can only be accomplished by adding *ad hoc* axioms.[25] When formal axiomatic systems, such as some mathematical theories modeling the physical universe, are supposed to represent a "final" or "universal" law of nature, they are not amenable to further rational analysis; besides, they are incomplete in Gödel's and Chaitin's sense, in that an infinity of their perfectly well-constructed propositions cannot be deduced by them. They can only be justified by their "success," defined as the capacity of such a system to generate a set of empirically testable predictions reasonably coincident with actual experimentally measured quantities, as illustrated in Figure 3.18. Any justification of some formal axiomatic system, beyond this operative "success," requires a wider meta-system, capable in its turn of outputting the system in question. Such meta-systems are, among others, all sorts of *ad hoc* omnipotent divinities. Consistency requires that this variety of meta-systems be nothing but irreducibly random contraptions; since *they* are, by definition, incompressible, one must accept the utter impossibility of their rational analysis: no logical language can *speak* about them, and they are ineffable. Moreover, there seems to be no acceptable alternative in sight, since the underpinning structure of Algorithmic Information Theory is shared by elementary mathematics; hence, to contest the validity of its startling conclusions is equivalent to contesting the validity of mathematics and logic, sapping the very foundations of our most fundamental tools of rational analysis.[25]

The random character of axiomatic systems sets a limit to the reduction of our knowledge of reality to some primordial, simple and hopefully compact mathematical explanation of nature, the dream of some contemporary physicists. A more modest ideal may consist in splitting the description of the (infinite? continuous?) universe into a finite number of partial algorithmically compressible fragments. In this scenario, our knowledge of the universe reduces to a finite set of axiomatic propositions, selected exclusively in view of their "success," but unassailable by any further analysis. These partial fragments remain algorithmically independent: the *joint information I(X, Y)* of the theories "explaining" two segments X and Y is $I(X) + I(Y)$, and their *mutual information content* $I(X:Y)$ shows that to know X does not help at all to calculate Y. A disturbing consequence of this independence is that, in this scenario, the sciences describing different portions of the universe may be based on mutually contradictory statements, and the consistency of the scientific corpus may be lost.

This, then, is the answer to the question: what do we know? if Algorithmic Information Theory is accepted as a valid tool for this analysis. And we emphasize: to reject this approach on the ground of preconceived convictions is to question mathematics (and logic) itself. All the same, let us not forget that Gödel's theorem, proving that arithmetic is *incomplete*, dates only from 1931. If such recent work has now shown beyond doubt that the set of primordial axioms founding our preeminent tool of analysis remains *uncertain*, then it may not be too surprising to state that the method first introduced in the *Timaeus* and adopted by all modern scientists after Galileo, the method underlying the Big Bang model, may also now found to be *limited*, that at most, only a partial, fragmentary knowledge is attainable. This frontier, here only sketched and vaguely perceived, can indeed illustrate the limits of *what we possibly can and what we effectively do know.*

But are the assertions deriving from Algorithmic Information Theory really so unconventional, have they in fact never been maintained in the course of history? The contrary seems to be the case: certain texts by Aristotle present a surprising homology with the conclusions just enunciated. We end our book by glancing at Aristotle's *Posterior Analytics*.

The Status of the Axioms in Aristotle's *Posterior Analytics*

Acquisition of knowledge is dealt with in different parts of the extensive Aristotelian corpus. To avoid misunderstandings, we first outline the *method* we will adhere to in the following interpretation of

Aristotle's *Posterior Analytics*.

1.   We refrain from reading into Aristotle's texts concepts and intentions which only modern investigation has developed, but do not reject the use of such modern concepts and meanings as instruments of elucidation of the same texts, insofar as they allow a finer analysis of his thoughts.

2.   We also refrain from settling in any way apparent or real discords detected by some authors in Aristotle's doctrine.

3.   We consider Aristotle's investigations in the *Posterior Analytics* to be a significant contribution to the analysis of the problem of scientific knowledge, a contribution still providing valuable insights to a contemporary reader.

4.   Passages quoted are for the most part very short excerpts, chosen because they represent some particular aspect of the problem. The only merit of this rather fragmentary approach is to avoid commentaries of unwieldy length.

We begin by briefly recasting Aristotle's investigations in the light of Algorithmic Information Theory.

In the two *Analytics*, Aristotle studies the conditions for the proper acquisition of scientific knowledge. "In the *Prior Analytics* Aristotle has stated and developed his theory of syllogism ... . He now (in the *Posterior Analytics*) turns to the problem of knowledge: what it is, how it is acquired, how guaranteed to be true, how expanded and systematized."[26]

First it must be clearly settled that: "Scientific knowledge cannot be acquired by sense-perception" (*Posterior Analytics* I 31, 87 b 28, cf. also I 88 a 9); and that: scientific knowledge is gathered by applying the syllogistic method, that is, by logically deducing (mechanically, with a Turing machine, for instance) consequences from certain first premises. But are these first premises true? Either they in turn are deduced from other first premises, and this then goes on *ad infinitum*, or such axioms are to be given a particular status. "Axioms," "first principles," "primary premises," are Aristotle's terms, which are homologous to "axiomatic formal systems," while "deduction" is what a Turing machine performs. A Turing machine's "list of states" is homologous to the rules of a correct syllogistic system. And "conclusions" or "theorems" is what scientific knowledge is all about.

Taken together, these deliberations show that if the initial premises are true, and "the notion[s] reached by demonstrative knowledge are also necessarily true" (*Posterior Analytics* I 4, 73 a 22), then, inescapably, the status of the axioms in Aristotle's theory of knowledge becomes crucial.

The opening sentence of the *Posterior Analytics* focuses directly on the problem outlined: "All teaching (*didaskalia*) and learning (*mathesis*) that involves the use of reason proceeds from preexistent knowledge" (*Posterior Analytics* I 1, 71 a 1-2). This previous knowledge may relate to the existence of some thing or fact, or to the definition or meaning of some term. He goes on: "We consider we have unqualified knowledge of anything when we believe that we know (i) ... the cause (*aitia*) from which the fact results, and (ii) that the fact cannot be otherwise" (*Posterior Analytics* I 2, 71 b 9-12). Science, that is, knowledge gleaned from the deductive method, has to ascertain that the cause attributed to a fact is truly its cause and must necessarily produce that fact as its effect.

This necessity is explicitly acknowledged by Aristotle: "If knowledge is such as we have assumed, demonstrative knowledge must proceed from premises which are true, primary, immediate, better known than, prior to, and causative of the conclusion" (*Posterior Analytics* I 2, 71 b 19-23). He goes on to list the characteristics of the premises: "The premises, then, must be true statements; because it is impossible to know that which is contrary to fact, e.g., that the diagonal of a square is commensurate with the sides. They must be primary and indemonstrable, because otherwise we shall not know them unless we have proofs of them" (*Posterior Analytics* I 2, 71 b 25-28). It follows that primary premises and principles must be identical notions: "To argue from primary premises is to argue from appropriate first principles (*arche*); for by primary premise and first principle I mean the same thing" (*Posterior Analytics* I 2, 72 a 5-7). This leads Aristotle to speak of the axioms: "I apply the term 'thesis' to an

immediate indemonstrable first principle of syllogism the grasp of
which is not necessary for the acquisition of certain kinds of
knowledge; but that [first principle] which must be grasped if any
knowledge is to be acquired, I call an 'axiom' (*axioma*)" (*Posterior
Analytics* I 2, 72 a 15-17). Aristotle thus defines *axiom as a
indemonstrable first principle absolutely indispensable for the
acquisition of scientific knowledge.* Thus, if the first principles are
indemonstrable, and if everything that can claim to be a scientific
knowledge is nothing but a deduction from these first principles, then
no scientific knowledge can be acquired unless these first principles
are somehow uncovered through other means.

    Aristotle readily admits this predicament. "We, however, hold
that not all knowledge is demonstrative; the knowledge of immediate
premises is not by demonstration" (*Posterior Analytics* I 3, 72 b 18-
20). Next a key result, fundamental for all scientific endeavor, is
propounded. If such an (*a priori* true) axiom system is "universal"
(*katholikos*), then the consequences deduced from it must be eternal,
i.e., universal laws of nature: "It is also evident that if the premises of
the syllogism are universal, the conclusion of a demonstration ... in the
strict sense, must be eternal" (*Posterior Analytics* I 8, 75 b 21-24).
Aristotle goes on to say: "It is difficult to be certain whether one
knows or not; for it is difficult to be certain whether our knowledge is
based upon the principles appropriate to each case – *it is this that
constitutes true knowledge* – or not" (*Posterior Analytics* I 9, 76 a 26-
28, our emphasis).

    In term of Algorithmic Information Theory, Aristotle's insights
can be reformulated as shown:

    Algorithmic compression: "It may be assumed that, given the
same conditions, that form of demonstration [i.e., science] is superior
to the rest which depends upon fewer postulates or hypotheses or
premises." (*Posterior Analytics* I 25, 86 a 33-35)

    Algorithmic information content: "The clearest indication that the
universal demonstration is more authoritative is that when we
comprehend the former of ... two premises we have knowledge in a
sense of the latter as well, and comprehend it potentially" (*Posterior*

Analytics I 24, 86 a 22 -25). When some system possesses universal algorithmic information content, it contains any particular system "potentially."

Algorithmic independence: "One science is different from another if their principles do not belong to the same genus, or if the principles of the one are not derived from the principles of the other" (*Posterior Analytics* I 28 87a 39 b1). Algorithmic independence is assimilated by Aristotle to the impossibility of the *metabasis eis allo genos*.[27] And he states: "Nor can a proposition of one science be proved by another science, except when the relation is such that the propositions of the one are subordinate to those of the other" (*Posterior Analytics* I 7, 75b 10-15).

But what about the possibility of knowing the axioms themselves? As noted by H. Tredennick, the translator of *Posterior Analytics*:

> There remains the question for whose answer Aristotle has repeatedly – by dramatic instinct – whetted our appetite: How do we apprehend the first principles themselves, which are not susceptible of demonstration? Is it by scientific knowledge – the same kind of knowledge by which we cognize demonstrable facts – or by a different faculty? If it is by a different faculty, how is it acquired? Still dramatic, Aristotle postpones his climax by taking the second point first. ... The faculty by which we know the [axioms themselves], since it cannot be either science or inferior to science, must be the only other intellectual faculty that is infallible, viz., *nous* or intuition,[28] which supervenes upon our logical process as a direct vision of truth.[29]

In fact, in the last paragraph of the last page of *Posterior Analytics*, Aristotle writes:

> Now of the intellectual faculties that we use in the pursuit of truth some (e.g., scientific knowledge and intuition) are always

true, whereas others (e.g., opinion and calculation) admit falsity; and no other kind of knowledge except intuition is more accurate than scientific knowledge. ... *It follows that there can be no scientific knowledge of the first principles*; and since nothing can be more infallible than scientific knowledge except intuition, it must be intuition that apprehends the first principles. This is also evident not only from the foregoing considerations but also because *the starting point of a demonstration is not itself a demonstration, and the starting point of scientific knowledge is not itself scientific knowledge"* (*Posterior Analytics* II 19, 100 b 5-17; our emphasis).

Resorting to the *nous* is Aristotle's own way of answering the question: how do we acquire knowledge of the first principles, if they cannot be deduced from any more fundamental assumptions?

Scientific knowledge has been found to be the outcome of a logical deductive process, starting from certain premises or axioms. In so far as these axioms are assumed to be true, the "mechanical" deductive procedure arrives at true conclusions, with certainty, that is, with probability equal one. But no science can say anything about its axioms, about these premises; in this sense, they remain ineffable. But neither Plato nor Aristotle accept this limit. For Plato as well as for Aristotle, the ultimate objective of philosophical endeavor is not the procurement of scientific knowledge, but the acquisition of "true" knowledge (*episteme*), which can only be knowledge of the *first* causes.

Their objective was to find *episteme*, the science by which the first causes of all reality could be uncovered. Their aim was truth (*aletheia*), not certainty. But they found that this aim was not attainable either by science, experience, or faith. The sole road leading to but never arriving at *episteme* and *aletheia* is the road pursued by the philosopher: the unrelenting, ceaseless re-posing of the questions *worth* asking.

In this way, the greatest philosophical systems and Algorithmic Information Theory converge on the same "tragic" – in the Greek sense – trait of the human condition: the Sisyphean everlasting pursuit of this *episteme*. But, nothing prevents us from "imagining a happy Sisyphus," as A. Camus so poignantly envisioned.[30]

Final Conclusions

If scientific knowledge is that kind of knowledge that rests on a mathematical theory, to what then, finally, does this kind of knowledge amount? Is there a clear-cut answer to this question? Or is it only possible to establish the *limits* of any pursuit of scientific knowledge? We have attempted to examine in what terms this problem is analyzed by Plato in the *Timaeus*, by Aristotle in *Posterior Analytics*, by contemporary Big Bang cosmology, and by Algorithmic Information Theory.

This analysis has shown the following conclusion to be unavoidable:

1) Scientific knowledge ultimately rests on a set of axiomatic propositions, propositions which however elude any subsequent scientific analysis.

2) No "mechanical" method leads *unerringly* to these propositions; they can only be uncovered by intuition, insight, or inspired guesses, of course after hard work. Thus no *method* for scientific discovery exists.

3) The validation of these axiomatic propositions is purely operative; they are retained if the model that can be achieved by their use is found to be adequate, if "it works", the way the Street Plan of New York adequately models ("speaks about"[31]) the infinite complexity of the city of New York.

4) The list of axiomatic propositions contains all the information of the theory admitting such axioms.

5) Questions about these propositions are meaningless within the theory.

6) Each such question is tantamount to an additional axiom.

7) No list of axioms can ever be claimed definitive, no "mechanical" method can answer the question: is this theory the

definitive one or can more primordial axioms be found?

However, the two most disturbing conclusions resulting from this research are:

8)   If *a priori* sweeping statements such as a pre-established harmony are ruled out, no theory can ever be in a position to compress the complexity of the universe into manageable propositions.

9)   About a theory that "does work" nothing meaningful, except that "it works," can be said; the theory can only be enunciated or copied.

"That whereof one cannot speak, thereof one must be silent."[32] Wittgenstein's final statement in his *Tractatus Logico-Philosophicus* finds application here, because we are finally led to conclude that

10) It is impossible to "speak scientifically" about scientific knowledge. In this sense, scientific knowledge reduces to ... silence.

Questioning the foundations of its most successful and undisputed activity, mankind encounters his finiteness. No one has so clearly seen in recent times the inherent *absurdity* of this permanent knocking one's head against clearly perceived unsurpassable limits as Albert Camus, who stated precisely that the outstanding characteristic of *the absurd consists in reason lucidly acknowledging its own limits.*[33]

NOTES

1. The text we translated is: "Wir Heutigen sind durch die eigentümliche Vorherr-
   schaft der neuzeitlichen Wissenschaften in den seltsamen Irrtum verstrickt, der
   meint, das Wissen lasse sich aus der Wissenschaft gewinnen und das Denken
   unterstehe der Gerichtsbarkeit der Wissenschaft. Aber das Einzige, was jeweils ein
   Denker zu sagen vermag, lässt sich logisch oder empirich weder beweisen noch
   widerlegen. Es ist auch nicht die Sache eines Glaubens. Es lässt sich nur fragend
   denkend zu Gesicht bringen. Das Gesichtete erscheint dabei stets als das
   Frag*würdige*." M. Heidegger, "Wer ist Nietzsches Zarathustra?" *Vorträge und
   Aufsätze*, Pfullingen 1954, p. 118-119.

2. In contrast, Kant admits as a valid source of knowledge the purely rational (i.e.,
   sense-independent) ascertainment of *a priori synthetic judgments*, directly
   prescribing universal laws to the physical world.

3. This method is at present the exclusive privilege of the physical sciences, such as
   cosmology. Biology, including genetics, as well as most human sciences, can – so
   far – only hope to achieve a similar status sometime in the future; at present, their
   predictions are at best "statistically significant," never universal "laws" of nature.

4. Some key aspects of the measuring operation, as pointed out by David Bohm in
   Wholeness and the Implicate Order, London 1980, deserve to be mentioned: 1)
   The intimate relation of the concept of measure with the concept of limit or
   boundary: "In the modern usage of the word 'measure,' the aspect of quantitative
   proportion or numerical ratio tends to be emphasized much more heavily than it
   was in ancient times. Yet even here the notion of boundary or limit is still present,
   though in the background. Thus, to set up a scale (e.g., of length) one must
   establish divisions which are in effect limits or boundaries of ordered segments" (p.
   119). 2) A concept deeper, more fundamental, and much more complex than
   measure is the concept of order: "The notion of order is so vast and so immense in
   its implications that it cannot be defined in words" (p. 115). 3) The notion of order
   similarly presupposes the concept of limit: "It is significant to note that in ancient
   times the most basic meaning of the word 'measure' was 'limit' or 'boundary'" (p.
   118). Indeed, the Greek word *metron* means "measure" and also "the point where
   the measure is completed," the "end" or "aim," i.e., a limit. 4) Measuring as
   performed in experiments, presupposes limits of (stable) objects and some
   hierarchy of ordered structures. But, "it is important to add here that order is not to
   be identified with predictability. *Predictability is a special kind of order such that
   a few steps determine the whole order*" (p. 118, our emphasis).

5. Cf. Kant's *Beharrlichkeit*, Plato's *summetria*. This prejudice also explains the uncomfortable feeling provoked by the idea of a dynamic, changing universe.

6. For instance, a *reductio ad absurdum* is accepted as a valid proof.

7. *If* such a mechanical procedure exists. But for all formal systems of interest, no such procedure is available! (see below)

8. Gregory J. Chaitin, *Algorithmic Information Theory*, Cambridge 1987; *Information, Randomness & Incompleteness. Papers on Algorithmic Information Theory*, Singapore 1987.

9. The following paragraphs are based on several works by Gregory J. Chaitin, with particular emphasis on his article "Gödel's Theorem and Information," *International Journal of Theoretical Physics* 22, 1982, p. 941-954, reprinted in *Information, Randomness & Incompleteness*, pp. 55-65.

10. Cf. Gregory J. Chaitin, "Information-theoretic Limitations of Formal Systems," *Journal of the Association for Computing Machinery* 21, 1974, p. 403-424, reprinted in *Information, Randomness & Incompleteness*, p. 165-190.

11. Gödel's theorem can also be expressed as follows: a formal axiomatic system sufficiently complex to be able to deal with elementary number theory, that is, with 1, 2, 3, 4, 5, ... plus addition and multiplication, can not be complete, even if it can be shown to be consistent: there will always remain some undecidable valid propositions which neither are true nor false. It follows that it is impossible to establish the internal logical consistency of a very large class of deductive systems – arithmetic, for example – and so no irreproachable warranty can be given that all of mathematics is entirely free from internal contradictions. This famous theorem ended the efforts of Whitehead and Russell in their *Principia Mathematica*, attempting to show that mathematics is reducible to logic, and put an end to the ambitious program of D. Hilbert which aimed at treating all of mathematics as a formal, self-contained axiom system. It was presented in 1931 in *Monatshefte für Mathematik und Physik* under the title "Über formal unentscheidbare Sätze der *Principia Mathematica* und verwandter Systeme I." At the time, Gödel was 25 years old.

12. Equivalently, a set of effective measurements or observations correlating to possible predictions, i.e., theorems, of the theory. We followed this method when we analyzed what all possible ideal cosmological measurements can tell us about the universe; cf. the second part.

13. *bit*, standing for "binary digit", is the elementary unit of information, it corresponds to the maximum information that one symbol can transmit. Generally, in binary notation, two symbols are used: 0 and 1.

14. To calculate π, here are some formulas:

$\pi/4 = 1/1 - 1/3 + 1/5 - 1/7 + \ldots$, or

$\pi/2 = \int_0^\infty dx/1 + x^2$.

15. Cf. J.E. Hopcroft, "Turing Machines," *Scientific American*, May 1984, p. 86-98, and J.L. Gersting, *Mathematical Structures for Computer Science*, New York 1983, chapter 9.

16. Cf. Gregory J. Chaitin, "A Theory of Program Size Formally Identical to Information Theory," *Journal of the Association for Computing Machines* 22, 1975, p. 329-340, reprinted in *Information, Randomness & Incompleteness*, Singapore 1987, p. 107-122.

17. Let us prove this seemingly incredible statement. Suppose one asks: how many n-bits sequences are generated by input programs not longer than $n - 10$ bits? By simply adding all the possible strings smaller than $n - 10$ bits, we obtain that there exist no more than $2^1 + 2^2 + \ldots + 2^{n-11}$ programs of length less than $n - 10$ bits. This sum does not exceed 2^{n-10}. It follows that fewer than 2^{n-10} programs have length less than $n - 10$ bits. These 2^{n-10} programs generate at most 2^{n-10} sequences. But there exist 2^n sequences n-bits long, and so one concludes that programs of length $n - 10$ bits only account for about one n-bit sequence in every thousand (since $2^{n-10} / 2^n = 1 / 1024$). This then shows that even such a modest compression as that involving only 10 bits is highly improbable, simply because the number of *possible* strings is so much greater than the number of compressed input strings that may output them. Thus, unless a pre-established symmetry is postulated, the probability of an *arbitrary* long but finite string being algorithmically random approaches certainty.

18. Similarly, Aristotle's *Organon* proposes an algorithm capable of deciding mechanically (hence "*organon*", that is, instrument) whether some proposition appertaining to the formal axiomatic system "the common language" is true or false, and moreover specifies the rules showing how to apply it to all possible cases. As J. Montserrat-Torrents puts it: "Inasmuch as a genetic kinship connects mind, language and reality, the study of the logical structure of the linguistic

expressions referring to reality can give valid information about the rational structure of the physical world." (J. Montserrat-Torrents, "Epistemological notes on Greek cosmologies", in *Foundations of Big Bang Cosmology*, F. Walter Meyerstein ed., Singapore 1989, p. 7). Thus it may not be too remote from the truth to conclude that the N-bits size of the formal system represented by the ancient Greek language, its algorithmic complexity, is measured by the N-bits of length of the *Organon*. This in turn measures *all* that such a formal system can *say* about the physical world. However, please carefully note that we have equated language with formal system. We are not unaware of the fact that the possibilities of a language allow it to transcend completely its strict formal constraints; poetry is one example. But, until further notice, no *scientific* knowledge of the physical world derives from poetry.

19. A cosmological theory should intend, at least as its ultimate objective, to explain *all* such observations as can be collected. But there are billions of stars in our galaxy, and billions of galaxies, and so on. (Cf. the second part.)

20. In this procedure, it is tacitly assumed that the segmentation of the undoubtedly gigantic, or maybe even infinite, sequence truly representing the universe, can be done without irreversible loss of information about the universe. Following once more the Platonic approach as outlined in the *Timaeus*, where this assumption is clearly accepted, such *a priori* supposition is the fundamental tenet on which the entire idea of reductionism ultimately rests.

21. If the machine halts, that is, if the calculation is successful.

22. R. Geroch and J.B. Hartle, "Computability and Physical Theories," in *Between Quantum and Cosmos*, W.H. Zurek *et al.* ed., Princeton N.J. 1988, pp. 549-566.

23. In *IBM Research Report* RC 19324, dated Dec. 12th, 1993, Gregory J. Chaitin, in a remarkable development, presents the software for a Universal Turing Machine that is easy to program and runs very quickly. This provides a new foundation for Algorithmic Information Theory. Now it is possible to write down executable programs that embody the constructions in the proofs of the theorems. Algorithmic Information Theory goes, in this way, from dealing with idealized *Gedankenexperimente* to being a theory about practical down-to-earth machines that one can actually construct and work with.

24. Quoted in S. Weinberg, *Dreams of a Final Theory*, New York, 1992, p. 115.

25. Beliefs, intimate convictions, prejudices, etc., are unaffected by these conclusions.

26. H. Tredennick, Introduction to Aristotle, *Posterior Analytics*, translated and commented by H. T., Loeb Classical Library, Cambridge Mass./London, 1960, p. 5.

27. *Metaphysics* V 28, 1024 b 14-15; X 4, 1055 a 6-7; X 7, 1057 a 26-28; XIII 9, 1085 a 16-19, 35 - b 4; *Prior Analytics* I 30, 46 a 17 ff.; *Topica* VIII 12, 162b 5-7; *De caelo* I 1, 268a 30 - b3.

28. *Intuition*, noun: Immediate apprehension by the mind without reasoning. "The Concise Oxford Dictionary".

29. H. Tredennick, Introduction to Aristotle, *Posterior analytics*, translated and commented by H. T., Loeb Classical Library, Cambridge Mass./London, 1960, p. 17.

30. "Il faut imaginer Sisyphe heureux", Albert Camus, *Le mythe de Sisyphe* [1942], Paris 1964, p. 166.

31. "The laws of physics, with all their logical apparatus, nevertheless speak, however indirectly, about the objects of the world" (L. Wittgenstein, *Tractatus Logico-Philosophicus* [1921, 1922], 6.3431, transl. by D.F. Pears & B.F. McGuinness, London 1961); in a similar manner, the Street Plan of New York "speaks" about New York.

32. "Worüber man nicht sprechen kann, darüber muss man schweigen", L. Wittgenstein, *Tractatus Logico-Philosophicus*, prop. 7.

33. Albert Camus, *Le mythe de Sisyphe*, "L'absurde, c'est la raison lucide qui constate ses limites", p. 70.

INDEX OF PROPER NAMES

GENERAL INDEX

Absurd, 180
Age of the universe, 106, 113, 119
Aitia, 21
Algorithm, 161, 164
Anagke, 22, 23, 88
Analogia, 27-28
Anthropic principle, 123, 169
Ausgewogenheit, 146 n. 71
Axiom, 1-4, 150-153, 153-155, 173-178, 139 n. 10

Baryon, 125
Being / Becoming 10
Big Bang [standard model], *passim*
Big Crunch, 106, 119
Bit, 183 n. 13
Black-body, cf. Planck
Black hole, 142 n. 32
Boson, 126
Boundary [of the universe], 79-80

Cauchy [surface of], 97
Cause, cf. *Aitia*
    causality, causal explanation, 20-21, 86-87
    causal stability condition, 95-96
CERN [Geneva], 126
Chaitin's probability, 160
Change, cf. *Kinesis*
Circle, 64 n. 8
Complexity,
    algorithmic, 157-162
    factor, 43
    of the sensible world, 166

Components, elementary, 41-48
Compression, algorithmic, 167-169
Computer, cf. Turing Machine
Connected, 139 n. 15
Consistency, logical, 182 n. 11
Content
    algorithmic information, 155, 157
    mutual information, 155
Contraction, cf. Expansion, 105-106
Copy, cf. *Eikon*
Cosmological or Copernican principle, 101
Cosmology, 67-68
Curvature, cf. Spacetime

Decoupling [radiation/matter], 108-109
Demiurge,
    not omnipotent, 22
    benevolent, 21
Density [critical/present], 119-120
Dependence, domain of, 97
Different, cf. Same
    Circle, 36
    Kind, 30
    Component of the world soul, 31
Differential equations, 80
Dimensions, of a manifold [number of], 80-81
Distance function, 83-84
Doppler effect, 143 n. 44
Doxa, cf. *Episteme*

Eidos, cf. *Idea*, 64 n. 1
Eikon, cf. Copy, 24